Progress in Probability and Statistics
Vol. 7

Edited by
Peter Huber
Murray Rosenblatt

Birkhäuser
Boston · Basel · Stuttgart

Seminar on Stochastic Processes, 1983

E. Çınlar
K. L. Chung
R. K. Getoor
editors

1984

Birkhäuser
Boston · Basel · Stuttgart

Editors:

E. Çınlar
Technological Institute
Northwestern University
Evanston, Illinois 60201

R. K. Getoor
Department of Mathematics
University of California—San Diego
La Jolla, California 92093

K. L. Chung
Department of Mathematics
Stanford University
Stanford, California 94305

CIP-Kurztitelaufnahme der Deutschen Bibliothek

Seminar on Stochastic Processes:
Seminar on Stochastic Processes . . . –
Boston ; Basel ; Stuttgart : Birkhäuser
3. 1983 (1984).
(Progress in probability and statistics ;
Vol. 7)
ISBN 3-7643-3293-X (Basel . . .)
ISBN 0-8176-3293-X (Boston . . .)

NE: GT

Printed in Germany
ISBN 0-8176-3293-X
ISBN 3-7643-3293-X

9 8 7 6 5 4 3 2 1

TABLE OF CONTENTS

FOREWORD

This volume consists of about half of the papers presented during a three-day seminar on stochastic processes. The seminar was the third of such yearly seminars aimed at bringing together a small group of researchers to discuss their current work in an informal atmosphere. The previous two seminars were held at Northwestern University, Evanston. This one was held at the University of Florida, Gainesville.

The invited participants in the seminar were B. ATKINSON, K.L. CHUNG, C. DELLACHERIE, J.L. DOOB, E.B. DYNKIN, N. FALKNER, R.K. GETOOR, J. GLOVER, T. JEULIN, H. KASPI, T. McCONNELL, J. MITRO, E. PERKINS, Z. POP-STOJANOVIC, M. RAO, L.C.G. ROGERS, P. SALMINEN, M.J. SHARPE, S.R.S. VARADHAN, and J. WALSH. We thank them and the other participants for the lively atmosphere they have created.

The seminar was made possible through the generous supports of the University of Florida, Department of Mathematics, and the Air Force Office of Scientific Research, Grant No. 82-0189, to Northwestern University. We are grateful for their support. Finally, we thank Professors Zoran POP-STOJANOVIC and Murali RAO for their time, effort, and kind hospitality in the organization of the seminar and during our stay in Gainesville.

E.Ç.
Evanston, 1984

Seminar on Stochastic Processes, 1983
Birkhäuser, Boston, 1984

A GENERAL THEORY APPROACH TO THE CONSTRUCTION OF MARKOV PROCESSES*

by

BRUCE W. ATKINSON

1. Introduction

This can be considered as the third in a series of papers exploiting the commutativity of projections for Markov processes as begun in [1] and continued in [2]. We use the projections here to address the problem of finding necessary and sufficient conditions for the existence of "very regular" Markov processes, which, among other things, serve to provide further insight into the familiar regularity assumptions of Markov process theory.

The starting point is a family $(\mathcal{F}(I))$ of sub-σ-fields of a probability space $(\Omega,\mathcal{F},P)$, where $I \subset (0,\infty)$ is an open interval. The family $(\mathcal{F}(I))$ can be thought of as a generalized stochastic process; this notion is attributed to Kolmogorov in [5]. The main question is: under what assumptions does there exist a right or left continuous Markov process with values in a "nice" state space so that $\mathcal{F}(I) = \bigvee_{t\in I} \sigma(X_t)$ almost surely. Even if $\mathcal{F}(I)$ were generated by some process, say Y, then we are asking that X be defined on the same probability space as

*This research was done while the author was employed by the University of Southern California.

Y. In other words, this is not a matter of simply checking finite dimensional distributions.

Before describing our method, we must point out that this same basic problem has been approached before in different contexts. In [5], Dynkin not only assumes the $\mathcal{F}(I)$ (which, together with a few additional assumptions to ours, he calls a stochastic system) but also assumes the existence of a "Markov representation" which has transition and co-transition probabilities. He then proceeds to prove the existence of a "regular Markov representation" with the same transition and co-transition probabilities.

Also, our technique is similar to the theory of prediction processes developed by Knight in [7]. Indeed, if Ω is the space of, say, right continuous paths ω with left limits from $\mathbb{R}^+$ to a compact metric space and $\mathcal{F}(I) = \sigma\{\omega(t);\ t \in I\}$, then the process X we construct is the analog of the prediction process. X is defined, via projection, as the collection of predictions on the past of processes adapted to the future.

The variations from the prediction process approach are two-fold. Firstly, we do not assume the existence of an underlying process which generates the $\mathcal{F}(I)$. Indeed, our first task here is to investigate intrinsic conditions on the $\mathcal{F}(I)$ which imply the existence of such an underlying process. Secondly, in determining the properties of the process constructed here, it is useful to not only construct X, which gives the predictions on the past of processes adapted to the future, but also to construct the process X_- (the left limit process of X) which turns out to give the reverse predictions on the future of processes adapted to the past. In other words we handle both time directions simultaneously. The pair (X_-, X) is shown to be strong Markov in both time directions.

We now briefly describe our method. We assume that, up to null sets, $\mathcal{F}(0,\infty)$ is countably generated and that $\mathcal{F}(I)$ is generated by the right hand germ fields $\mathcal{F}_{(t]} = \bigcap_{t<u} \mathcal{F}(t,u)$, $t \in I$. These are obvious necessary conditions for the existence of a right continuous process which generates the $\mathcal{F}(I)$. Our Markovian assumption is the germ-Markov property: $\bigcap_{t<u} \mathcal{F}(0,u)$ and $\bigcap_{s<t} \mathcal{F}(s,\infty)$ are conditionally independent given $\mathcal{F}_{[t]} = \bigcap_{s<t<u} \mathcal{F}(s,u)$. We note that in connection with prediction processes, Knight has analyzed the germ-Markov property in [6].

In section 2 we construct a countable family $Y = (Y^n)$ of continuous processes which generate all measurable processes. In section 3 we apply the optional projection followed by the reverse predictable projection to the processes Y^n to get a process $V = (V^n)$; V corresponds to the prediction process. Then we apply the reverse optional projection followed by the predictable projection to the processes Y^n to get a process $U = (U^n)$; U corresponds to the reverse prediction process. It is shown that V is right continuous with left limits and U is left continuous with right limits. Finally we formally define $X = (U_+, V)$ and $X_- = (U, V_-)$; see (3.5).

In section 4 we show that, due to the germ-Markov property, X generates the processes that are optional and reverse predictable, X_- generates the processes that are predictable and reverse optional, and $\mathcal{F}(I) = \bigvee_{t\in I} \sigma(X_t) = \bigvee_{t\in I} \sigma(X_{t-})$; see (4.1). Theorem (4.3) states that the familiar σ-fields associated with random times are defined quite naturally in terms of the paths of X. For example, the past up to a random time R is generated by $X_{R\wedge t}$, $t \in [0,\infty]$. Finally (4.4) states that X is strong Markov and reverse moderately Markov, X_- is moderately Markov and reverse strong Markov, and (X_-, X) is strong Markov in both time directions.

Section 5 concludes with a few remarks and applications. We make some specific comments on how X can be regarded as the regularization of the $\mathcal{F}(I)$. For example, if the $\mathcal{F}(I)$ are initially generated by a right continuous process which is Markov relative to its right continuous natural filtration, then even if this process does not have left limits, or is not strong Markov, we can "replace" it by one that has these properties. We end with analogs of Revuz measures and Motoo's theorem.

2. Hypotheses and a Few Easy Consequences

Let $(\Omega,\mathcal{F},P)$ be a probability space. For each open interval $I \subset (0,\infty)$ let $\mathcal{F}(I)$ be a sub-σ-field of $\mathcal{F}$ such that $\mathcal{F}(0,\infty) = \mathcal{F}$ and each $\mathcal{F}(I)$ contains all P-null sets of $\mathcal{F}$. (Note: $\mathcal{F}(a,b)$ will stand for the more cumbersome $\mathcal{F}((a,b))$. Also, since in this paper the probability P is fixed, a "null set" shall always mean a P-null set of $\mathcal{F}$.)

(2.1) DEFINITIONS

(a) If $0 < t \le \infty$, $\mathcal{F}_{t-} \equiv \mathcal{F}(0,t)$; $\mathcal{F}_{0-} \equiv \bigcap_{0<t} \mathcal{F}_{t-}$.

(b) If $0 \le t < \infty$, $\mathcal{F}_t \equiv \bigcap_{t<u} \mathcal{F}_{u-}$; $\mathcal{F}_\infty \equiv \mathcal{F}$.

(c) If $0 \le t < \infty$, $\hat{\mathcal{F}}_{t+} \equiv \mathcal{F}(t,\infty)$; $\hat{\mathcal{F}}_{\infty+} \equiv \bigcap_{t<\infty} \hat{\mathcal{F}}_{t+}$.

(d) If $0 < t \le \infty$, $\hat{\mathcal{F}}_t \equiv \bigcap_{s<t} \hat{\mathcal{F}}_{s+}$; $\hat{\mathcal{F}}_0 \equiv \mathcal{F}$.

(e) If $0 < t \le \infty$, $\mathcal{F}_{[t)} \equiv \bigcap_{s<t} \mathcal{F}(s,t)$; $\mathcal{F}_{[0)} \equiv \mathcal{F}_0$.

(f) If $0 \le t < \infty$, $\mathcal{F}_{(t]} \equiv \bigcap_{t<u} \mathcal{F}(t,u)$; $\mathcal{F}_{(\infty]} \equiv \hat{\mathcal{F}}_\infty$.

(g) If $0 < t < \infty$, $\mathcal{F}_{[t]} \equiv \bigcap_{s<t<u} \mathcal{F}(s,u)$; $\mathcal{F}_{[0]} \equiv \mathcal{F}_0$; $\mathcal{F}_{[\infty]} \equiv \mathcal{F}_\infty$.

Observe that $\mathcal{F}_{0-} = \mathcal{F}_0 = \mathcal{F}_{[0)} = \mathcal{F}_{(0]} = \mathcal{F}_{[0]}$ and $\hat{\mathcal{F}}_{\infty+} = \hat{\mathcal{F}}_\infty = \mathcal{F}_{[\infty)} = \mathcal{F}_{(\infty]} = \mathcal{F}_{[\infty]}$ by definition. Also compare (2.1) with (2.1) of [2].

(2.2) HYPOTHESES

(a) For every open interval $I \subset (0,\infty)$, $\mathcal{F}(I) = \bigvee_{t\in I} \mathcal{F}_{(t]}$.

(b) $\mathcal{F}$ is *countably generated up to null sets;* in other words, there is a countable set $\{f_n : n \in \mathbb{N}\} \subset b\mathcal{F}$ such that

$$\mathcal{F} = \sigma(f_n : n \in N) \vee (\text{null sets}).$$

(c) For every $t \in [0,\infty]$, $\mathcal{F}_t$ and $\hat{\mathcal{F}}_t$ are conditionally independent (hereafter abbreviated as c.i.) given $\mathcal{F}_{[t]}$.

(2.3) REMARKS. (a) Hypotheses (2.2a,b) are necessary conditions for there to exist a right continuous process X with values in a compact metric space such that $\mathcal{F}(I) = \bigvee_{t\in I} \sigma(X_t)$ almost surely. Actually there is no reason, in our theory, to prefer right continuity to left continuity. Indeed, it will be shown later that, in the presence of (2.2b,c), (2.2a) is equivalent to having $\mathcal{F}(I) = \bigvee_{t\in I} \mathcal{F}_{[t)}$ almost surely for every open interval $I \subset (0,\infty)$; see (5.3).

(b) Hypothesis (2.2c) represents the appropriate Markovian assumption. Since $\mathcal{F}_{[t]}$ is a two-sided germ field, (2.2c) is a two-sided Markov property. However, this property is equivalent to each of the one-sided Markov properties:

(i) $\mathcal{F}_{t-}$ and $\hat{\mathcal{F}}_t$ are c.i. given $\mathcal{F}_{[t)}$ for every t.

(ii) $\mathcal{F}_t$ and $\hat{\mathcal{F}}_{t+}$ are c.i. given $\mathcal{F}_{(t]}$ for every t.

This follows from Theorem 2.3 of [2] where, for the purpose of [2], $\mathcal{G}_t \equiv \mathcal{F}_{(t]}$.

(c) There are many examples of systems which satisfy (2.2). Let $(X_t : 0 \le t \le \infty)$ be a stochastic process with values in a compact metric space. Assume as given a probability P such that

(i) $\mathcal{F} = [\bigvee_{0\le t\le\infty} \sigma(X_t)] \vee$ (null sets),

(ii) the paths of X are P-almost surely right continuous, and

(iii) X is Markov relative to P in the sense that $\bigvee_{s\leq t} \sigma(X_s)$ and $\bigvee_{t\leq u} \sigma(X_u)$ are c.i. given X_t for every $0 \leq t \leq \infty$.

For each open interval $I \subset (0,\infty)$, define $\mathcal{F}(I) = [\bigvee_{t\in I} \sigma(X_t)] \vee$ (null sets). It follows that the system $\mathcal{F}(I)$ satisfies (2.2); see [1], section 2.

Actually, the main purpose of this paper is to show that, in the proper sense (to be seen), this is the only example which gives rise to (2.2). Furthermore, we shall see that we can also impose a number of very nice properties on the process X we construct; see section 4.

(2.4) DEFINITIONS. (All real valued processes in this paper are defined on $[0,\infty] \times \Omega$.)

(a) A real valued process Z is said to be *evanescent* if $\{\omega: Z_t(\omega) \neq 0 \text{ for some } t\}$ is null. The set of evanescent processes will be denoted by $\mathcal{I}$.

(b) Let $\mathfrak{M} \equiv (\mathcal{B}[0,\infty] \otimes \mathcal{F}) \vee \mathcal{I}$. If Z is $\mathfrak{M}$-measurable we simply call Z *measurable*.

For ease in using the monotone class theorem, we can assume, without loss of generality, that the collection $\{f_n: n \in \mathbb{N}\}$ is closed under multiplication and $f_1 = 0$, $f_2 = 1$, $0 \leq f_n \leq 1$; see (2.2b) above.

Letting $\mathcal{S} \equiv \sigma(f_n: n \in N)$ we have $\mathcal{F} = \mathcal{S} \vee$ (null sets). An easy application of the monotone class theorem yields:

(2.5) PROPOSITION. $\mathfrak{M} = (\mathcal{B}[0,\infty] \otimes \mathcal{S}) \vee \mathcal{I}$.

For $n \in \mathbb{N}$, let $g_n: [0,\infty] \to [0,1]$ be continuous and suppose $g_1 = 0$ and $g_2 = 1$. Further suppose that $\{g_n: n \in \mathbb{N}\}$ is closed under multiplication and $\sigma(g_n: n \in \mathbb{N}) = \mathcal{B}[0,\infty]$.

Let $Y^1, Y^2, \dots$ be an enumeration of all processes of the form $Y_t(\omega) = g_m(t)f_n(\omega)$ for $m,n \in \mathbb{N}$. Using (2.5), the monotone class theorem, and the fact that $\{Y^n: n \in N\}$ is closed under multiplication we have the following.

(2.6) PROPOSITION. (a) $\sigma(Y^n: n \in N) = \mathcal{B}[0,\infty] \otimes \mathcal{S}$.

(b) $\mathfrak{M} = \sigma(Y^n: n \in N) \vee \mathcal{I}$.

3. Construction of X

(3.1) DEFINITIONS. $\mathcal{P}$ (resp. $\mathcal{O}$, $\hat{\mathcal{P}}$, $\hat{\mathcal{O}}$) is the σ-field on $[0,\infty] \times \Omega$ generated by real valued processes Z such that Z_t is measurable relative to $\mathcal{F}_{t-}$ (resp. $\mathcal{F}_t$, $\hat{\mathcal{F}}_{t+}$, $\hat{\mathcal{F}}_t$) and $t \to Z_t$ is left (resp. right, right, left) continuous P-a.s. A process Z which is measurable relative to $\mathcal{P}$ (resp. $\mathcal{O}$, $\hat{\mathcal{P}}$, $\hat{\mathcal{O}}$) is called *predictable* (resp. *optional*, *co-predictable*, *co-optional*); see [1]-3.1).

For $Z \in b\mathfrak{M}$, ${}^pZ$ (resp. ${}^oZ$, ${}^{\hat{p}}Z$, ${}^{\hat{o}}Z$) $\in b\mathcal{P}$ (resp. $\mathcal{O}$, $\hat{\mathcal{P}}$, $\hat{\mathcal{O}}$) is the predictable (resp. optional, co-predictable, co-optional) projection of Z. ${}^pZ$ and ${}^oZ$ are defined relative to the filtration $(\mathcal{F}_t)$ while ${}^{\hat{p}}Z$ and ${}^{\hat{o}}Z$ are defined relative to the reverse filtration $(\hat{\mathcal{F}}_t)$ via time reversal; see [1]-section 3.

We will need the following path regularity property which follows from [4]-VI, 47-50, appropriately interpreted for the co-optional and co-predictable projections. First we establish some terminology.

(3.2) DEFINITIONS. Let M be a metric space and $h: [0,\infty] \to M$.

(a) h is *r.c.l.l.* means h is right continuous for $t \in [0,\infty)$ and has left limits for $t \in (0,\infty]$.

(b) h is *l.c.r.l.* means h is left continuous for $t \in (0,\infty]$ and has right limits for $t \in [0,\infty)$.

(3.3) PROPOSITION. (a) *Suppose* $W \in b\mathfrak{M}$ *and that* $t \to W_t$ *is r.c.l.l.* P-*a.s. Then so is any version of* ${}^{o}W$ *or* ${}^{\hat{p}}W$.

(b) *Suppose* $W \in b\mathfrak{M}$ *and that* $t \to W_t$ *is l.c.r.l.* P-*a.s. Then so is any version of* ${}^{p}W$ *or* ${}^{\hat{o}}W$.

We now apply this result to the processes Y^n; see (2.6). Since Y^n is both r.c.l.l. and l.c.r.l. then (3.3), together with [1] - Theorem (3.2), implies that there exists U^n (resp. V^n) $\in b(\mathcal{P} \cap \hat{\mathcal{O}})$ (resp. $b(\mathcal{O} \cap \hat{\mathcal{P}})$) with values in [0,1] such that U^n (resp. V^n) is a version of ${}^{p\hat{o}}Y^n$ (resp. ${}^{o\hat{p}}Y^n$) and $t \to U^n_t$ (resp. V^n_t) is l.c.r.l. (resp. r.c.l.l.) for all ω.

NOTE: Here, e.g., ${}^{p\hat{o}}Y^n \equiv {}^{p}({}^{\hat{o}}Y^n)$. Also we use the word version in the sense that any two versions must differ by an evanescent process. See [1]-Section 4 for an intrinsic characterization of the iterated projections $Z \to {}^{p\hat{o}}Z$ and $Z \to {}^{o\hat{p}}Z$.

(3.4) THEOREM. (a) $\mathcal{P} \cap \hat{\mathcal{O}} = \sigma(U^n\colon n \in N) \vee \mathcal{I}$.

(b) $\mathcal{O} \cap \hat{\mathcal{P}} = \sigma(V^n\colon n \in N) \vee \mathcal{I}$.

PROOF. We only show (a) as (b) is similar.

By the preceding discussion we already know that $\sigma(U^n\colon n \in N) \vee \mathcal{I} \subset \mathcal{P} \cap \hat{\mathcal{O}}$.

To prove the opposite inclusion let $\mathcal{H} = \{W \in b\mathfrak{M}\colon \exists$ a version W^1 of ${}^{p\hat{o}}W$ such that $W^1 \in b\sigma(U^n\colon n \in N)\}$. It is easy to check that $\mathcal{H}$ is a monotone vector space. Also, $\{Y^n\colon n \in N\} \cup \mathcal{I} \subset \mathcal{H}$. Thus the monotone class theorem implies, together with (2.6b), that $\mathcal{H} = b\mathfrak{M}$.

Now suppose $W \in b(\mathcal{P} \cap \hat{\mathcal{O}})$. Since W is a version of ${}^{p\hat{o}}W$, then there exists $W^1 \in b\sigma(U^n\colon n \in N)$ such that $W - W^1$ is evanescent. That is to say that $W \in b(\sigma(U^n\colon n \in N) \vee \mathcal{I})$, which is what we needed.

(3.5) DEFINITION. (a) U^n_+ is the process such that $(U^n_+)_t = U^n_{t+}$ for $0 \le t < \infty$, and $(U^n_+)_\infty = U^n_\infty$.

(b) V^n_- is the process such that $(V^n_-)_t = V^n_{t-}$ for $0 < t \le \infty$, and $(V^n_-)_0 = V^n_0$.

(c) $X = (X^1, X^2, \ldots)$ is the process with values in $[0,1]^N$ so that for $n \in N$, $X^{2n-1} = U^n_+$ and $X^{2n} = V^n$.

(d) $X_- = (X^1_-, X^2_-, \ldots)$ is the process with values in $[0,1]^N$ so that for $n \in N$, $X^{2n-1}_- = U^n$ and $X^{2n}_- = V^n_-$.

The next result is immediate from the properties of U^n and V^n.

(3.6) PROPOSITION. X *is r.c.l.l.,* X_- *is l.c.r.l.,* $(X_-)_t = X_{t-}$ *for* $0 < t \le \infty$, *and* $X_t = (X_-)_{t+}$ *for* $0 \le t < \infty$.

4. Properties of X

In the previous section we gave the path regularity properties of X and X_-. The properties in this section are divided between further regularity properties and Markov properties.

(4.1) THEOREM. (a) $\mathcal{P} \cap \hat{\mathcal{O}} = \sigma(X_-) \vee \mathcal{I}$, $\mathcal{O} \cap \hat{\mathcal{P}} = \sigma(X) \vee \mathcal{I}$, *and* $\mathcal{O} \cap \hat{\mathcal{O}} = \sigma(X_-,X) \vee \mathcal{I}$.

(b) *For every* t, $\mathcal{F}_{[t)} = \sigma((X_-)_t) \vee$ *(null sets),* $\mathcal{F}_{(t]} = \sigma(X_t) \vee$ *(null sets), and* $\mathcal{F}_{[t]} = \sigma((X_-)_t, X_t) \vee$ *(null sets).*

(c) *For every open interval* $I \subset (0,\infty)$, $\mathcal{F}(I) = [\bigvee_{t\in I} \sigma((X_-)_t)] \vee$ *(null sets)* $= [\bigvee_{t\in I} \sigma(X_t)] \vee$ *(null sets).*

PROOF. By (3.4b) and the definition of X, in order to prove that $\mathcal{O} \cap \hat{\mathcal{P}} = \sigma(X) \vee \mathcal{I}$ it is enough to check that $U^n_+ \in b(\mathcal{O} \cap \hat{\mathcal{P}})$ for $n \in N$.

Since $U^n \in b(\mathcal{P} \cap \hat{\mathcal{O}})$ it follows that $U^n_t \in b\mathcal{F}_{[t)}$. This is a consequence of the left hand Markov property; see (2.3b). It is evident that for $0 \le t < \infty$, $U^n_{t+} \in b\mathcal{F}_{(t]}$. Also $(U^n_+)_\infty \in b\mathcal{F}_{[\infty)} = b\mathcal{F}_{(\infty]}$. By (3.1), $U^n_+ \in b(\mathcal{O} \cap \hat{\mathcal{P}})$. Thus $\mathcal{O} \cap \hat{\mathcal{P}} = \sigma(X) \vee \mathcal{I}$.

The proof that $\mathcal{P} \cap \hat{\mathcal{O}} = \sigma(X_-) \vee \mathcal{I}$ is similar.

It will be convenient to prove parts of (4.1b) and (4.1c) before completing the proof of (4.1a). By definition of projections and due to the right and left Markov properties (see (2.3b)) we have that $\mathcal{F}_{[t)} = \sigma(W_t\colon W \in b(\mathcal{P} \cap \hat{\mathcal{O}})$, and $\mathcal{F}_{(t]} = \sigma(W_t\colon W \in b(\mathcal{O} \cap \hat{\mathcal{P}}))$. By what was just shown, $\mathcal{F}_{[t)} = \sigma((X_-)_t) \vee$ (null sets) and $\mathcal{F}_{(t]} = \sigma(X_t) \vee$ (null sets). This gives the first two parts of (4.1b). Also (4.1c) now follows from (2.2a) and (3.6).

There now remains to prove the last parts of (4.1a) and (4.1b). Since X is r.c.l.l. and $\mathcal{F} = \left(\bigvee_{0<t<\infty} \sigma(X_t)\right) \vee$ (null sets), then the exact same proof as for [2]-(4.3) shows $\mathcal{M} = \mathcal{P} \vee \hat{\mathcal{P}}$. Let $W = W^1W^2$, where $W^1 \in b\mathcal{P}$ and $W^2 \in b\hat{\mathcal{P}}$. Then ${}^{o\hat{o}}W = ({}^{\hat{o}}W^1)({}^{o}W^2)$. But by [1]-(3.2) ${}^{\hat{o}}W^1 \in b(\mathcal{P} \cap \hat{\mathcal{O}})$ and ${}^{\hat{o}}W^2 \in b(\mathcal{O} \cap \hat{\mathcal{P}})$. The monotone class theorem implies that $\mathcal{O} \cap \hat{\mathcal{O}} = (\mathcal{P} \cap \hat{\mathcal{O}}) \vee (\mathcal{O} \cap \hat{\mathcal{P}})$ since such W generate $\mathcal{M}$ and since the optional and co-optional projections commute; see [1]-(3.8). The third part of (4.1a) now follows from the first two, and since $\mathcal{F}_{[t]} = \sigma(W_t\colon W \in b(\mathcal{O} \cap \hat{\mathcal{O}})$ (a consequence of (2.2c)), then the third part of (4.1b) is also evident.

Before stating Markov properties it is necessary to discuss random times.

(4.2) DEFINITIONS. (a) A *random time* R is a $[0,\infty]$ valued random variable defined on $(\Omega,\mathcal{F})$.

(b) R is *optional* if for every t, $\{R < t\} \in \mathcal{F}_t$.

(c) R is *predictable* if R is optional and $\{(t,\omega): t = R(\omega)\} \in \mathcal{P}$.

(d) R is *reverse optional* if for every t, $\{R > t\} \in \hat{\mathcal{F}}_t$.

(e) R is *reverse predictable* if R is reverse optional and $\{(t,\omega): t = R(\omega)\} \in \hat{\mathcal{P}}$.

(f) $\mathcal{F}_{R-} \equiv \sigma(W_R: W \in b\mathcal{P})$.

(g) $\mathcal{F}_R \equiv \sigma(W_R: W \in b\mathcal{O})$.

(h) $\hat{\mathcal{F}}_{R+} \equiv \sigma(W_R: W \in b\hat{\mathcal{P}})$.

(i) $\hat{\mathcal{F}}_R \equiv \sigma(W_R: W \in b\hat{\mathcal{O}})$.

(4.3) THEOREM. *Let* R *be a random time.*

(a) $\mathcal{F}_{R-} = [\bigvee_t \sigma((X_-)_{R \wedge t})] \vee$ (null sets).

(b) $\mathcal{F}_R = [\bigvee_t \sigma(X_{R \wedge t})] \vee$ (null sets).

(c) $\hat{\mathcal{F}}_{R+} = [\bigvee_t \sigma(X_{R \vee t})] \vee$ (null sets).

(d) $\mathcal{F}_R = [\bigvee_t \sigma((X_-)_{R \vee t})] \vee$ (null sets).

PROOF: (a) Since X_- is l.c.r.l. and $(X_-)_t$ is $\mathcal{F}_{[t)}$-measurable it follows easily that $[\bigvee_t \sigma((X_-)_{R \wedge t})] \vee$ (null sets) $\subset \mathcal{F}_{R-}$.

We now go through a few cases to reverse that inclusion. First, let $f \in b\mathcal{B}([0,1]^N)$ and let $W_t = 1_{\{0\}}(t)f((X_-)_0)$. Then $W_R = f((X_-)_{R \wedge 0})1_{\{R=0\}}$. It follows by (4.1b) that for every $F \in b\mathcal{F}_{0-}$, $W_R \in b([\bigvee_t \sigma((X_-)_{R \wedge t})] \vee \sigma(R) \vee$ (null sets), where $W_t = 1_{\{0\}}(t)F$. However, observe that if $h \in b\mathcal{B}([0,\infty])$, then the deterministic process $h(t)$ is in $b(\mathcal{P} \cap \hat{\mathcal{O}})$. Thus there exists, by (4.1a), $f \in b\mathcal{B}([0,1]^N)$ so that $h(t) - f((X_-)_t)$ is evanescent. Therefore, $h(R) = f((X_-)_R) = f(X_-)_{R \wedge \infty})$ P-a.s., and $\sigma(R) \subset [\bigvee_t \sigma((X_-)_{R \wedge t})] \vee$ (null sets). Thus for W as above, $W_R \in b[\bigvee_t \sigma((X_-)_{R \wedge t})] \vee$ (null sets).

Next, let $0 < a < b < \infty$ and $f \in b\mathcal{B}([0,1]^N)$. Letting $W_t =$

$1_{(b,\infty]}(t)f(X_{a-})$ we have that $W_R = f((X_-)_{R\wedge a})1_{\{b<R\}}$. Since by (4.1c), $\mathcal{F}_{b-} = [\bigvee_{0<t<b} \sigma((X_-)_t)] \vee$ (null sets) then the monotone class theorem implies $W_R \in b[(\bigvee_t \sigma((X_-)_{R\wedge t})) \vee \text{(null sets)}]$ for W of the form $W_t = 1_{(b,\infty]}F$, where $F \in b\mathcal{F}_{b-}$. (Note that again we have used the discussion in the preceding paragraph concerning the measurability of R.)

Since the two types of processes W just discussed must generate $\mathcal{P}$, the monotone class theorem implies $\mathcal{F}_{R-} \subset [\bigvee_t \sigma((X_-)_{R\wedge t})] \vee$ (null sets).

(b) Since X is r.c.l.l. and X_t is $\mathcal{F}_{(t]}$-measurable, then $[\bigvee_t \sigma(X_{R\wedge t})] \vee \text{(null sets)} \subset \mathcal{F}_R$.

For the other inclusion we observe that if $W^1 \in b\mathcal{P}$ and $W^2 \in b\hat{\mathcal{P}}$, then ${}^o(W^1W^2) = [(W^1)\cdot({}^oW^2)] \in b(\mathcal{P} \vee (\mathcal{O} \cap \hat{\mathcal{P}})) = b(\mathcal{P} \vee \sigma(X))$. Since, as was used in the proof of (4.1), $\mathcal{M} = \mathcal{P} \vee \hat{\mathcal{P}}$, the monotone class theorem implies $\mathcal{O} = \mathcal{P} \vee \sigma(X)$.

Using the fact that $\mathcal{F}_{0-} = \sigma(X_0) \vee$ (null sets) and $\mathcal{F}_{b-} = [\bigvee_{0<t<b} \sigma(X_t)] \vee$ (null sets) (for $0 < b < \infty$), and modifying slightly the proof of (a) above, it follows that for $W \in b\mathcal{P}$, $W_R \in b[(\bigvee_t \sigma(X_{R\wedge t})) \vee \text{(null sets)}]$. Also, if $f \in b\mathcal{B}([0,1]^N)$ then $f(X_R) = f(X_{R\wedge\infty}) \in b[(\bigvee_t \sigma(X_{R\wedge t})) \vee \text{(null sets)}]$. As a result of the preceding paragraph we have $\mathcal{F}_R \subset [\bigvee_t \sigma(X_{R\wedge t})] \vee$ (null sets), which is the desired result.

(c) and (d) are proved similarly. They are the "dual" results of (a) and (b), respectively, and can be obtained via time reversal.

(4.4) THEOREM. *Let* R *be a random time.*

(a) *If* R *is predictable or reverse optional then* $\mathcal{F}_{R-}$ *and* $\hat{\mathcal{F}}_R$ *are* c.i. *given* $(X_-)_R$.

(b) *If* R *is optional or reverse predictable then* $\mathcal{F}_R$ *and* $\hat{\mathcal{F}}_{R+}$ *are* c.i. *given* X_R.

(c) *If* R *is optional or reverse optional then* $\mathcal{F}_R$ *and* $\hat{\mathcal{F}}_R$ *are* c.i. *given* $(X_-)_R$, X_R.

PROOF. We will prove (b) only as (a) and (c) are quite similar.

Let R be optional and $W \in b\hat{\mathcal{P}}$. Then, by definition, $E(W_R|\mathcal{F}_R) = {}^oW_R$. By [1]-(3.2), ${}^oW \in b(\mathcal{O} \cap \hat{\mathcal{P}})$ and by (4.1a), ${}^oW_R \in b(\sigma(X_R) \vee \text{(null sets)})$. Hence $\mathcal{F}_R$ and $\hat{\mathcal{F}}_{R+}$ are c.i. given X_R.

Next, let R be reverse predictable and $W \in b\mathcal{O}$. By definition, $E(W_R|\hat{\mathcal{F}}_{R+}) = {}^{\hat{p}}W_R$. Again by [1]-(3.2), ${}^{p}W \in b(\mathcal{O} \cap \hat{\mathcal{P}})$ and by (4.1a), ${}^{\hat{p}}W_R \in b(\sigma(X_R) \vee \text{(null sets)})$. Thus $\mathcal{F}_R$ and $\hat{\mathcal{F}}_{R+}$ are c.i. given X_R.

(4.5) REMARKS. (a) (4.4a) is a path property of (X_-). Indeed, combined with (4.3a,d) it says that whenever R is predictable or reverse optional then the pre-R X_- process and the post-R X_- process are c.i. given $(X_-)_R$.

(b) Similarly if R is optional or reverse predictable then the pre-R X process and the post-R X process are c.i. given X_R.

(c) In the same vein, if R is optional or reverse optional then the pre-R X process and the post-R X_- process are c.i. given $(X_-)_R$, X_R. Actually, it is easy to check that if $Z = (X_-, X)$, then (4.4c) becomes a path property of Z: the pre-R Z process and the post-R Z process are c.i. given Z_R.

5. Observations

(5.1) REGULARITY OF FILTRATIONS. Consider a fixed random time, $R = t$. Applying (4.3b) we see that $\mathcal{F}_t = \left(\bigvee_{s\le t} \sigma(X_s)\right) \vee \text{(null sets)}$. In other words, the natural completed filtration corresponding to X is right continuous. Similarly (4.3a) implies $\mathcal{F}_{t-} = \left(\bigvee_{s\le t} \sigma((X_-)_s)\right) \vee$ (null sets), and the natural completed filtration for X_- is left continuous. (Incidentally, it also follows that for $t > 0$, $\mathcal{F}_{t-} = \left(\bigvee_{s<t} \sigma(X_s)\right) \vee$ (null sets).) Reversing time we get $\hat{\mathcal{F}}_t = \left(\bigvee_{s\ge t} \sigma((X_-)_s)\right) \vee$ (null sets)

and $\hat{\mathcal{F}}_{t+} = \left(\bigvee_{s\geq t} \sigma(X_s)\right) \vee$ (null sets).

(5.2) MODERATE MARKOV PROPERTY OF X_-. (4.4a) implies that X_- is a left continuous moderate Markov process; see [3].

(5.3) SYMMETRY OF HYPOTHESES. Combining (4.1b) and (4.1c) we see that for every open interval $I \subset (0,\infty)$, $\mathcal{F}(I) = \bigvee_{t\in I} \mathcal{F}_{[t)}$. Now, if we use this property to replace (2.2a), it follows that, in a similar fashion, (2.2a) would be implied. The condition that $\mathcal{F}(I) = \bigvee_{t\in I} \mathcal{F}_{[t)}$ is obviously necessary for there to exist a left continuous process generating the $\mathcal{F}(I)$. Thus, in the presence of (2.2b,c), the "right continuity hypothesis," (2.2a), is equivalent to the "left continuity hypothesis," i.e. that $\mathcal{F}(I) = \bigvee_{t\in I} \mathcal{F}_{[t)}$.

(5.4) SPECIAL CASES. (a) $\mathcal{F}(I)$ *generated by a filtration.* Suppose $(\Omega,\mathcal{F},P)$ is a probability space and $(\mathcal{H}_t: 0 < t < \infty)$ is a right continuous filtration contained in $\mathcal{F}$ such that each $\mathcal{H}_t$ contains all P-null sets relative to $\mathcal{F}$ and $\mathcal{F} = \bigvee_{0<t<\infty} \mathcal{H}_t$. For $I \subset (0,\infty)$, an open interval, let $\mathcal{F}(I) \equiv \bigvee_{t\in I} \mathcal{H}_t$.

Obviously, if $0 < t < \infty$, then $\mathcal{F}_{(t]} = \mathcal{H}_t$, and condition (2.2a) is satisfied. Since, for $0 < t < \infty$, $\mathcal{F}_{[t]} = \mathcal{F}_t = \mathcal{H}_t$, (2.2c) is also satisfied. Thus (2.2b) is the only one of three that has content for this situation, and we make it an assumption. Observe also that $\hat{\mathcal{F}}_{t+} = \hat{\mathcal{F}}_t = \mathcal{F}$ and $\hat{\mathcal{O}} = \hat{\mathcal{P}} = \mathfrak{M}$. Now (4.1) implies that $\mathcal{P} = \sigma(X_-) \vee \mathcal{I}$, $\mathcal{O} = \sigma(X) \vee \mathcal{I}$, $\mathcal{F}_{t-} = \sigma((X_-)_t) \vee$ (null sets), and $\mathcal{F}_t = \mathcal{H}_t = \sigma(X_t) \vee$ (null sets). Compare this with [8].

Note that in this case (4.4) has no content since the "past" equals the "present." Of course, a similar discussion is possible if the $\mathcal{F}(I)$ are generated by a reverse filtration.

(b) *Right continuous Markov processes.* As mentionned in (2.3c), if $\mathcal{F}(I) = \bigvee_{t\in I} \sigma(Y_t) \vee$ (null sets) where Y is a right continuous simple Markov process (i.e. as specified in (2.3c)), then the $\mathcal{F}(I)$ satisfy (2.2). Then X, as constructed in section 3, may be thought of as a strong Markov regularization of Y. However, in general it will only be true that $\sigma(X_t) = \bigcap_{t<s} \bigvee_{t<u<s} \sigma(Y_u)$ a.s., and hence for every open interval $I \subset (0,\infty)$, $\mathcal{F}(I) = \bigvee_{t\in I} \sigma(Y_t) = \bigvee_{t\in I} \sigma(X_t)$ a.s. But if we further assume that Y is Markov relative to the right continuous filtration, $(\mathcal{F}_t)$, i.e. $\mathcal{F}_t$ and $\bigvee_{u\geq t} \sigma(Y_u)$ are c.i. given Y_t, then $\sigma(X_t) = \sigma(Y_t)$ for every t.

Finally if we assume Y is right continuous and strong Markov in the sense that $\mathcal{F}_R$ and $\bigvee_t \sigma(Y_{t\vee R})$ are c.i. given Y_R for optional R, it then follows that $\sigma(X_R) = \sigma(Y_R)$ a.s. for optional R. Since X has left limits, then X can be considered a "compactification" of Y, albeit X may have a different state space. Note that in any case, even though there is no assumption of left limits for Y, "left limits" must exist in the sense that $\mathcal{F}(I) = \bigvee_{t\in I} \mathcal{F}_{[t)}$.

(5.5) APPLICATIONS TO "ADDITIVE FUNCTIONALS." A random measure A(dt), for $t \in [0,\infty]$, is said to be *optional and co-optional* if for every $0 < a < b < \infty$, $A((a,b))$ is $\mathcal{F}(a,b)$-measurable, $A(\{0\})$ is $\mathcal{F}_0$-measurable, and $A(\{\infty\})$ is $\mathcal{F}_\infty$-measurable. Also, A is called *finite* if $EA([0,\infty]) < \infty$. See [1] - Sections 3 and 4. Using these results we have:

(a) *The collection of all finite random measures* A *which are optional and co-optional is in one-to-one correspondence with the collection of finite measures* μ *on* $[0,1]^N \times [0,1]^N$ *which do not charge inaccessible sets. (Here,* $\Gamma \subset [0,1]^N \times [0,1]^N$ *is inaccessible if* $\{(t,\omega): ((X_-)(t,\omega), X(t,\omega)) \in \Gamma\}$ *is evanescent.) The correspondence obtains as follows: Let* $W \in b\mathfrak{M}$. *By* (4.1a) *there exists* $f \in b\mathcal{B}([0,1]^N \times [0,1]^N)$ *so that* $f(X_-,X) - {}^{o\hat{o}}W$ *is evanescent. Then*

$$E \int W_t \, A(dt) = \int f \, d\mu.$$

The measure μ, associated to A, is the analog of the Revuz measure associated to an additive functional.

(b) *If* A *and* B *are finite random measures which are optional and co-optional and* $A(dt) << B(dt)$ a.s., *then there exists* $f \in b\mathcal{B}([0,1]^N \times [0,1]^N)^+$ *so that*

$$A(dt) = f((X_-)_t, X_t) \, B(dt) \text{ a.s.}$$

This is the analog of Motoo's theorem.

References

1. B. ATKINSON. Generalized strong Markov properties and applications. *Z. Wahr. verw. Geb. 60* (1982), 71-78.

2. B. ATKINSON. Germ fields and a converse to the strong Markov property. *Seminar on Stochastic Processes* - 1982, pp. 1-21, Birkhäuser, Boston 1983.

3. K.L. CHUNG and J. GLOVER. Left continuous moderate Markov processes. *Z. Wahr. verw. Geb. 49* (1979), 237-248.

4. C. DELLACHERIE and P. MEYER. *Probabilités et Potentiel.* Herman, Paris, 1980.

5. E.B. DYNKIN. Markov representations of stochastic systems. *Russian Math. Surveys, 30* (1975), 65-104.

6. F. KNIGHT. Prediction processes and an autonomous germ Markov property. *Annals of Prob., 7* (1979), 385-405.

7. F. KNIGHT. *Essays on the Prediction Process.* IMS Lecture Note Series, Hayward, 1981.

8. P.A. MEYER and YEN KIA-AN. Génération d'une famille de tribus par un processus croissant. *Séminaire de Probabilités* IX, Springer-Verlag, Lecture Notes 465 (1975), 466-470.

Department of Mathematics
University of Florida
Gainesville, Florida 32611

Seminar on Stochastic Processes, 1983
Birkhäuser, Boston, 1984

CONDITIONAL GAUGES

by

K. L. CHUNG*

Let D be a bounded domain in R^d, $d \geq 1$; $x_0 \in D$; H the class of positive harmonic functions h with $h(x_0) = 1$. Let $\{X_t, t \geq 0\}$ be the Brownian motion in D killed at ∂D, with transition semigroup (P_t^D); $\tau = \tau_D$ the first exit time from D. Let q be a bounded Borel function, and

$$e(t) = e_q(t) = \exp\{\int_0^t q(X_s)ds\}.$$

For $h \in H$ let P_h^x and E_h^x denote the probability and expectation determined by the h-conditioned process ("h-process" for short) starting at x; when $h \equiv 1$, it is omitted from the notation. The key relation is as follows: if $Y \geq 0$ and $Y \in F_t$ (measurable w.r. to $F_t = \sigma(X_s, 0 \leq s \leq t)$, augmented as usual), we have

$$(1) \qquad E_h^x\{Y;\ t < \tau_D\} = \frac{1}{h(x)} E^x\{Y \cdot h(X_t);\ t < \tau_D\}.$$

We fix D and q in this paper, and denote the *gauge* for (D,q) by u (see [2]). This notation can be extended to any h-process if we define the h-gauge by

*Research supported in part by NSF grant MCS-83-01072 at Stanford University.

$$u_h(x) = E_h^x\{e_q(\tau_D)\}. \tag{2}$$

It is proved in [1] that either $u \equiv \infty$ in D or u is bounded in $\bar{D}$. In the latter case we shall say that "the gauge is finite." N. Falkner extended the result as follows.

THEOREM. *Suppose that the domain is Green-smooth. If for some* $h \in H$ *and some* $x \in D$ *we have* $u_h(x) < \infty$, *then*

$$\sup_{h \in H} \sup_{x \in D} u_h(x) < \infty. \tag{3}$$

Moreover, this is the case if and only if the gauge is finite.

In this note we give a simpler proof of this result by the methods used in [2], together with the smoothness properties developed in [3].

The definition of a "Green-smooth domain" is given in [3]. We need only the following consequences, where $G = G_D = \int_0^\infty P_t^D \, dt$ is the Green's operator for D.

(a) There exists C_1 such that for all $h \in H$: $\int_D h \, dm \le C_1$.

(b) There exists $C_2 > 0$ such that for all $h \in H$: $C_2 G1 \le h$.

(c) There exists $C_3 > 0$ such that for all $h \in H$: $C_3 Gh \le h$.

From now on we shall use a single symbol C to denote any strictly positive constant, which depends on D and q in general, and whose value may vary. Any other dependence will be indicated by parentheses. We use the following notation from [1]:

$$L_t f(x) = E^x\{t < \tau_D;\ e_q(t) f(X_t)\}$$

$$Vf(x) = \int_0^\infty dt\ L_t f(x).$$

It is proved in [1] and [2] that the following assertions are equivalent:

(i) the gauge is finite;

(ii) $V1 < \infty$;

(iii) $L_t 1 \le C\, e^{-Ct}$ for all $t \ge 0$.

PROOF OF THE THEOREM. Suppose the gauge is finite, and write Q for $\|q\|$. We have easily

$$L_t h \le e^{Qt}\, P_t^D h \le e^{Qt}\, h\,;$$

hence for each $a > 0$:

$$\int_0^a L_t h\, dt \le a\, e^{Qa}\, h\,. \tag{4}$$

Using (a), we have also

$$L_a h \le e^{Qa} \frac{1}{(2\pi a)^{d/2}} \int_D h\, dm \le C(a). \tag{5}$$

Let $R_a = \int_a^\infty L_t\, h\, dt$. Then by (iii), since $L_{a+t} h \le L_t(L_a h) \le \|L_a h\| L_t 1$:

$$R_a \le \|L_a h\| \int_0^\infty L_t 1\, dt \le C(a). \tag{6}$$

Here and hereafter $\|\cdots\|$ denotes the sup-norm for Borel functions. For $1 \le s \le 2$, we have

$$R_3 = L_s\, R_{3-s} \le L_s\, R_1 \le e^{2Q}\, P_s^D\, R_1. \tag{7}$$

Hence by (b) and (6):

$$R_3 \le e^{2Q} \int_1^2 P_s^D\, R_1\, ds \le e^{2Q}\, \|R_1\|\, G1 \le Ch. \tag{8}$$

It follows from (4) and (8) that

$$\int_0^\infty L_t\, h\, dt \le 3\, e^{3Q}\, h + Ch = Ch. \tag{9}$$

Since $e_q(t) \in F_t$, we see from (1) that after an application of Fubini's theorem:

$$E_h^x\{\int_0^\tau e_q(t)|q|(X_t)dt\} = \int_0^\infty dt\ E_h^x\{t < \tau;\ e_q(t)|q|(X_t)dt\}$$

$$\leq Q\frac{1}{h(x)}\int_0^\infty dt\ L_t h(x) \leq QC.$$

Replacing $|q|$ by q above and integrating, the first term becomes

$$E_h^x\{e_q(\tau) - 1\}.$$

Hence

$$u_h(x) = E_h^x\{e_q(\tau)\} \leq 1 + QC$$

and this yields (3). Thus the finiteness of the gauge implies all h-gauges are uniformly bounded.

Next, suppose for a fixed $h \in H$ and a fixed $x \in D$ we have $u_h(x) < \infty$. Thus for each $s > 0$:

$$\infty > \sum_{n=0}^{\infty} E_h^x\{ns < \tau \leq (n+1)s;\ e(\tau)\}. \tag{10}$$

Since the h-process is (strongly) Markovian, the expectation above is equal to

$$E_h^x\{ns < \tau;\ e(ns)E_h^{X(ns)}[0 < \tau \leq s;\ e(\tau)]\}. \tag{11}$$

By (c) applied only for the fixed h, we have

$$\sup_{x\in D} E_h^x(\tau_D) = \sup_{x\in D}\frac{1}{h(x)}\,Gh(x) \leq C. \tag{12}$$

Hence by Chebychev's inequality, for any $\delta < 1$ there exists $s > 0$ such that

$$\inf_{x\in D} P_h^x(\tau_D \leq s) > \delta. \tag{13}$$

It follows that

(14) $$\inf_{x\in D} E_h^x\{0 < \tau \le s;\ e(\tau)\} \ge \delta e^{-Qs}.$$

Substituting (14) into (11), we obtain from (10)

(15) $$\infty > \sum_{n=0}^{\infty} P_h^x\{ns < \tau;\ e(ns)\} = \sum_{n=0}^{\infty} E^x\{ns < \tau;\ e(ns)h(X(ns))\}$$

$$= \sum_{n=0}^{\infty} L_{ns}h(x).$$

It is well known that $\{h(X_t)1_{\{t<\tau\}}, F_t, P^x\}$ is a supermartingale. It follows that we have for $s < t$:

(16) $$E^x\{s < \tau;\ e(s)h(X(s))\} \ge E^x\{t < \tau;\ e(s)h(X(t))\}$$

$$\ge E^x\{t < \tau;\ e(t)h(X(t))\}e^{Q(s-t)}.$$

Hence if $ns \le t \le (n+1)s$, we have

(17) $$L_{ns}h(x) \ge e^{-Qs} L_t h(x) \ge e^{-2Qs} L_{(n+1)s}h(x)$$

and consequently (15) implies

(18) $$\infty > \int_0^{\infty} dt\ L_t h(x) = Vh(x).$$

(It is easy to see that $L_t h(x)$ is Borel measurable in t, in fact continuous.) Writing g for $G1$, we infer from (18) and (b) applied for the fixed h that

(19) $$V(|q|g) < \infty.$$

Observe that $g(x) = E^x\{\tau_D\}$ for $x \in D$, and that $\tau - t = \tau \circ \theta_t$ on

$\{t < \tau\}$. Hence we obtain by the Markov property:

$$E^x\{t < \tau;\ e(t)q(X_t)(\tau - t)\} = E^x\{t < \tau;\ e(t)q(X_t)g(X_t)\} = L_t(qg).$$

Since (19) holds, we can apply Fubini's theorem to obtain

$$\infty > E^x\{\int_0^\tau e(t)q(X_t)(\tau - t)dt\} = E^x\{e(t)(\tau - t)\big|_0^\tau + \int_0^\tau e(t)dt\}$$

$$= -E^x\{\tau\} + E^x\{\int_0^\tau e(t)dt\}.$$

Since $E^x(\tau) < \infty$, this implies

$$V1(x) = E^x\{\int_0^\tau e(t)dt\} < \infty. \tag{20}$$

This is equivalent to the finiteness of the gauge as reviewed above. The proof is so simple that we will give it here: by (20) and Fubini's theorem:

$$\infty > Vq(x) = E^x\{\int_0^\tau e_q(t)q(X_t)dt\} = E^x\{e_q(\tau_D) - 1\}.$$

References

1. K.L. CHUNG and K.M. RAO. Feynman-Kac functional and the Schrödinger equation. *Seminar on Stochastic Processes, 1981*, pp. 1-29. Birkhäuser, Boston, 1981.

2. K.L. CHUNG. An inequality for boundary value problems. *Seminar on Stochastic Processes 1982*, pp. 111-122. Birkhäuser, Boston, 1983.

3. N. FALKNER. Feynman-Kac functionals and positive solutions of $\frac{1}{2}\Delta u + qu = 0$. To appear in *Z. Wahrscheinlichkeitstheorie*.

K.L. Chung
Department of Mathematics
Stanford University
Stanford, CA 94305

Seminar on Stochastic Processes, 1983
Birkhäuser, Boston, 1984

DUALITY UNDER A NEW SETTING

by

K. L. Chung*, Ming Liao* and K. M. Rao

0. Introduction

This is a continuation of [2]. The developments there are complicated by an exceptional set denoted by Z (see [2], p. 179). It is shown that Z is a polar set under the conditions there if and only if Hunt's Hypothesis (B) holds (see [2], p, 192). In this paper a set of sufficcient conditions on the potential kernel will be given for the absence of Z. These strengthen the conditions used in [2]. The dual semigroup $\{\hat{P}_t, t \geq 0\}$ (see 2 , p. 191) is then defined on the state space E_∂, some of its properties will be reviewed and adduced. A process will then be constructed with the dual semigroup as its transition semigroup, which will be shown to be a Hunt process on E_∂. This process is in (strong) duality with the original Hunt Process in the sense of [1].

1. The Strengthened Hypotheses

Let $X = \{X_t, t \geq 0\}$ be a Hunt process with a locally compact, second countable state space E; and let $E_\partial = E \cup \{\partial\}$ be the one-point compactification of E. Let ξ be a Radon measure on $(E,\mathcal{E})$ where $\mathcal{E}$

*Research supported in part by NSF grant MCS83-01072.

is the Borel tribe on E. Let $\{P_t, t \geq 0\}$ be the transition semigroup, and U the potential kernel of X, such that

$$(1) \qquad U(x,dy) = u(x,y)\xi(dy) .$$

The conditions on u and ξ are listed below for ready reference. Throughout this paper, the symbol K is reserved for a compact subset of E.

(i) $u(x,y) > 0$ for all $(x,y) \in E\times E$; $u(x,y) = +\infty$ if and only if $x = y \in E$;

(ii) for all $x \in E$: $u(x,\cdot)$ is extended continuous in E;

(iii) for all $y \in E$: $u(\cdot,y)$ is extended continuous in E;

(iv) for all $y \in E$: $\lim_{x\to\partial} u(x,y) = 0$;

(v) there exists $x_o \in E$: $\lim_{y\to\partial} u(x_o,y) = 0$;

(vi) for all K and for all $x \in E$: $\int_K u(x,y)\xi(dy) < \infty$;

(vii) for all K and for all $y \in E$: $\int_K \xi(dx)u(x,y) < \infty$;

(viii) ξ is an excessive measure.

Not all these conditions will be needed in all the results below. Let us compare them with those used in [2]. In that paper only conditions (i), (ii) and an alternative form of (vi) are imposed until §5, where (viii) and another condition implied by (iii) are added. Thus the new hypotheses are (iii), (iv), (v) and (vii). Let us remark that for the purposes of this paper, condition (iii) may be weakened as follows:

(iii') for all $y \in E$: $u(\cdot,y)$ is lower semicontinuous and bounded off a neighborhood of y.

Note that (ii) and (iv) together are equivalent to the continuity of

$u(\cdot,y)$ in E_∂ with $u(\partial,y) = 0$, for each $y \in E$. On the other hand, (iii) and (v) together imply only $u(x_o,\cdot)$ is continuous in E_∂ and $u(x_o,\partial) = 0$, for some $x_o \in E$. We shall set $u(x,\partial) = 0$ for each $x \in E$, without necessarily assuming the continuity of $u(x,\cdot)$ at ∂. Thus $u(x,y)$ is now defined for $(x,y) \in E_\partial \times E_\partial$ except at (∂,∂); and is equal to zero if $x = \partial$ or $y = \partial$.

The condition (vi), namely $U(\cdot,K)$ is a finite function, implies that there exists $h \in \mathcal{E}_+$ such that $0 < Uh < \infty$ on E, which in the presence of (iii) implies that

(2) $$\lim_{t \to \infty} P_t P_K 1 = 0 \ .$$

The last property is used as an assumption in [2], which is referred to as the transience of X. Let us remark that under (iii) or (iii'), $u(\cdot,y)$ is excessive for each y, hence the $\underline{u}(\cdot,y)$ which appears in [2] is simply $u(\cdot,y)$ here. Finally, let us recall the definition of a "potential" ([2], p. 169) as an excessive function s such that

(3) $$\lim_{K \uparrow E} P_{K^c} s = 0, \qquad \xi\text{-a.e.}$$

A "pure potential" is a potential such that

(4) $$\lim_{t \to \infty} P_t s = 0 \qquad \xi\text{-a.e.}$$

The set Q is defined ([2], p. 190) as follows:

(5) $$Q = \{y \in E: \ u(\cdot,y) \text{ is a pure potential}\}.$$

It is proved in [2] (Theorem 10, p. 195) that $Q \subset Z^c$. Hence in order to prove $Z = \emptyset$ it is sufficient to prove $Q = E$. The proof below does not use condition (v) or (viii).

THEOREM 1. For each $y \in E$, $u(\cdot,y)$ is a pure potential.

PROOF. As remarked above, $u(\cdot,y)$ is excessive. We have

$$P_{K^c} u(x,y) = \int P_{K^c}(x,dz)u(z,y) . \tag{6}$$

As $P_{K^c}(x,dz)$ is concentrated on $\overline{K^c}$, while under condition (iv) $u(z,y)$ converges to zero as z leaves all compact subsets of E, it is clear that (3) is satisfied when $s = u(\cdot,y)$, indeed for all $x \in E$. Thus $u(\cdot,y)$ is a potential.

Fix a compact K and let V be open such that $y \in V \subset K^o$. For $t > 1$ we have

$$P_t u(x,y) \leq \int_{t-1}^{t} P_s u(x,y)ds = \int_{t-1}^{t} \left(\int_V + \int_{K-V} + \int_{E-K}\right) P_s(x,dz)u(z,y)ds . \tag{7}$$

If $x \neq y$, then $u(x,y) < \infty$; we can choose V so that

$$\sup_{z \in V} u(x,z) = M_1 < \infty$$

by (ii); and

$$\int_V \xi(dz)u(z,y) \leq \varepsilon$$

by (vii). It follows that

$$\int_{t-1}^{t} \int_V P_s(x,dz)u(z,y)ds \leq \int_V U(x,dz)u(z,y) = \int_V u(x,z)u(z,y)\xi(dz) \leq M_1\varepsilon.$$

Next, we have

$$\sup_{z \in E-V} u(z,y) \leq M_2 < \infty$$

by (iii); hence

$$\int_{t-1}^{t} \int_{K-V} P_s(x,dz)u(z,y)ds \leq \int_{K-V} P_{t-1} U(x,dz)u(z,y) \leq M_2 P_{t-1} U(x,K).$$

Since $U(x,K) < \infty$, the last term above converges to zero as $t \to \infty$. Lastly we can choose K so that

$$\sup_{z \in E-K} u(z,y) \le \varepsilon$$

by condition (iv). Hence

$$\int_{t-1}^{t} \int_{E-K} P_s(x,dz)u(z,y)ds \le \varepsilon .$$

Combining the estimates above, we obtain if $x \neq y$:

$$\lim_{t \to \infty} P_t\, u(x,y) = 0 . \tag{8}$$

Thus (4) is true when $s = u(\cdot,y)$ with the only exceptional set $\{y\}$. Since $\xi(\{y\}) = 0$, we have proved $u(\cdot,y)$ is a pure potential.

Q.E.D.

COROLLARY. For any measure μ on $\mathcal{E}$, if $U\mu \not\equiv \infty$, then $U\mu$ is a pure potential.

PROOF. Recall that even under the more general conditions of [2] (Corollary on p. 171) we have $U\mu < \infty$, ξ-a.e. Let $U\mu(x) < \infty$, then $\mu(\{x\}) = 0$ because $u(x,x) = \infty$; and $P_t\, U\mu(x) < \infty$, $P_{K^c}\, U\mu(x) < \infty$ because $U\mu$ is excessive. It follws that

$$\infty > P_t\, U\mu(x) = \int_{E-\{x\}} P_t\, u(x,y)\mu(dy) \downarrow 0$$

as $t \to \infty$, by (8); and

$$\infty > P_{K^c} U\mu(x) = \int_E P_{K^c}\, u(x,y)\mu(dy) \downarrow 0$$

as $K \uparrow E$, by the argument under (6).

We state some of the consequences of the result "$Z = \emptyset$", which we have just proved.

(a) For each $y \in E$, and open neighborhood G of y, we have

$$u(x,y) = P_G\, u(x,y), \qquad x \in E, \quad y \in G.$$

This is actually equivalent to "$Z = \emptyset$", see p. 173 of [2]; but we shall not use this directly.

(b) For any two measures μ and ν on $\mathcal{E}$, if $U\mu = U\nu \not\equiv \infty$ on E, then $\mu = \nu$, see p. 186 of [2].

(c) Riesz representation. If f is an excessive function $\not\equiv \infty$, then there exists a unique Radon measure μ and a harmonic function h such that

$$f = U\mu + h. \tag{9}$$

It is proved in Theorem 6, p. 187 of [2], under more general conditions, that we have such a representation with $\mu(Z) = 0$, and that $Q \subset Z^c$. But it does not follow that $U\mu$ is a potential since μ may change $Q^c - Z$. Under the present conditions, $U\mu$ is a potential by the Corollary to Theorem 1. In [2], the uniqueness of μ is proved only for the class of μ with $\mu(Z) = 0$, so that the possibility of another representation like (9) but with $\mu(Z) > 0$ cannot be ruled out. The absence of Z ensures the uniqueness of representation.

Finally, we will record a simplified form of the principal convergence theorem in [2]. This combines Theorem 2 and its continuation on p. 176 and p. 180 of [2], and corrects a slip there.

THEOREM 2. Suppose $Z = \emptyset$. Let $\{\mu_n\}$ be a sequence of measures on $\mathcal{E}$ satisfying the conditions:

(a) for all n: $U\mu_n \leq \sigma$, where σ is a potential;

(b) $\lim_n U\mu_n = s$, where s is an excessive function $\not\equiv \infty$.

Then $\{\mu_n\}$ converges vaguely to a Radon measure μ, and $s = U\mu$.

The word "diffuse" in the original statement of Theorem 2 should be deleted. The only place this superfluous (and inconvenient) condition was apparently used is in (2) on p. 177 of [2]. But since $u(x,x) = \infty$, $U\mu_n(x) < \infty$ implies $\mu_n(\{x\}) = 0$, which is what we need there. The vague convergence of the entire sequence $\{\mu_n\}$ follows from the unique determination of any subsequential vague limit, on account of consequence (b) above.

2. The Dual Semigroup

Recall the definition of the dual semigroup $\{\hat{P}_t, t \geq 0\}$ on p. 190 of [2]*. For each $y \in E$ and $t \geq 0$, $P_t u(\cdot,y)$ is a pure potential by Theorem 1 above. Hence by the Riesz theorem there is a unique measure μ such that $P_t u(\cdot,y) = U\mu$. This measure is denoted by $\hat{P}_t(\cdot,y)$. Thus we have for all x and y:

$$(1) \qquad P_t u(x,y) = \int P_t(x,dz)u(z,y) = \int u(x,z)\hat{P}_t(dz,y).$$

Let $g \in b\mathcal{E}_+$ with compact support, so that $Ug < \infty$ by condition (vi). We have by (1)

$$(2) \qquad \int u(x,z)\int \hat{P}_t(dz,y)g(y)\xi(dy) = \int P_t u(x,y)g(y)\xi(dy) = P_t Ug(x)$$

$$= UP_t g(x) = \int u(x,z)P_t g(z)\xi(dz).$$

It follows from uniqueness of the Riesz representation that

$$(3) \qquad \int \hat{P}_t(dz,y)g(y)\xi(dy) = P_t g(z)\xi(dz)$$

*It can be verified that $\{\hat{P}_t\}$ is Borelian.

as measures. For any $f \in b\mathcal{E}_+$, if we integrate f with respect to the two measures in (3), we obtain

$$(4) \qquad \int (f\hat{P}_t)g \, d\xi = \int f(P_t g)d\xi \, .$$

This is the duality relation between (P_t) and $(\hat{P}_t)$.

Since $Q = E$, the following results are proved in Theorem 7, p. 191, Theorem 8, p. 193, and (31), p. 195 of [2].

THEOREM 3. $\{\hat{P}_t, t \geq 0\}$ is a submarkovian semigroup on $E \times E$. For each y, $\hat{P}_t(\cdot,y)$ is weakly right continuous in $t \geq 0$, and $\hat{P}_o(\cdot,y) = \delta_y$. Furthermore, we have

$$(5) \qquad \hat{U}(dx,y) = \xi(dx)u(x,y) \, .$$

The weak right continuity is a consequence of vague right continuity, the semigroup property, and the weak right continuity at $t = 0$. The proof depends on Theorem 2 in §1. Under condition (vii), $\hat{U}(K,y) < \infty$ for each $y \in E$ and compact $K \subset E$.

We extend $(\hat{P}_t)$ to E_∂ to be strictly Markovian in the usual way. Namely, we set $\hat{P}_t(\{\partial\},\partial) = 1$ for $t \geq 0$; and $\hat{P}_t(\{\partial\},y) = 1 - \hat{P}_t(E,y)$ for $y \in E$, $t \geq 0$. We denote this extension by $(\hat{P}_t)$ without changing the notation.

3. Coexcessive Functions

A function on E is coexcessive iff f is excessive with respect to $(\hat{P}_t)$ on E. If we put $f(\partial) = 0$, this can be extended to E_∂. We shall do so in the sequel. Since $\hat{U}(K,\cdot) < \infty$ for each compact $K \subset E$, it follows from a general theorem ([3], p. 86) that: if f is coexcessive, then there exist $g_n \in b\mathcal{E}_+$ such that

$$(1) \qquad f = \lim_{n \to \infty} \uparrow g_n \hat{U} .$$

In fact we may choose $g_n \leq n^2$ and $g_n \hat{U} \leq n$. By condition (ii) and Fatou's lemma, $g_n \hat{U}$ is l.s.c. (lower semicontinuous). Hence each coexcessive function is l.s.c. by (1). It is a consequence of (1) and $u > 0$ that a coexcessive function $\not\equiv 0$ is strictly positive on E (cf. Proposition 9 on p. 171 of [2]). If f_1 and f_2 are both coexcessive, then $f_1 \wedge f_2$ is superaveraging with respect to $(\hat{P}_t)$, and l.s.c. Since $(\hat{P}_t)$ is weakly right continuous, it follows that $f_1 \wedge f_2$ is coexcessive. Similarly if f is coexcessive and c is a constant ≥ 0, then $f \wedge c$ is coexcessive. For each $x \in E$, $u(x,\cdot)$ is coexcessive by (1) of §2, because for each $y \in E$, $u(\cdot,y)$ is excessive.

Let $C_o(E_\partial)$ denote the class of continuous functions on E_∂ which vanish at ∂. We define

S = the class of coexcessive functions which belong to $C_o(E_\partial)$;

$L = S - S$.

By condition (v), $u(x_o,\cdot) \in S$. Indeed, we may replace that condition by the following:

(v') there exists a member of S which is not identically zero.

It will be seen that all the arguments below remain valid if we replace $u(x_o,\cdot)$ by any member $\not\equiv 0$ of S.

It seems of interest to examine the significance of condition (v). This has to do with the lifetime $\hat{\zeta}$ of the dual process. Let

$$\lambda(y) = \hat{P}^y\{\hat{\zeta} = \infty\} , \qquad y \in E_\partial .$$

Then λ is coexcessive, and it follows as in Proposition 9, p. 171 of [2] that if $\lambda \not\equiv 0$ in E, then $\lambda > 0$ in E. The following result can be proved in a way similar to that of Theorem 1.

PROPOSITION 4. Suppose $\lambda > 0$ in E. If for any $x \in E$, the function $u(x,\cdot)$ is continuous in E_∂, then $u(x,\partial) = 0$.

A dual proposition for $u(\cdot,y)$ is also true. Thus, suppose $P^x\{\zeta = \infty\} > 0$ for some (hence all) $x \in E$, and $u(\cdot,y)$ is purely excessive. If $u(\cdot,y)$ is continuous at ∂, then $u(\partial,y) = 0$.

From now on all the conditions in §1 will be used.

THEOREM 5. L is dense in $C_o(E_\partial)$ endowed with the sup-norm.

PROOF. S is a cone which is closed under the minimum operation, and also under truncation by a constant $c \geq 0$, as reviewed above. Let $x_1 \neq x_2$, both in E. Then since $u > 0$ on $E \times E$, there exists a constant $A > 0$ such that $Au(x_o,x_1) > u(x_1,x_2)$, where x_o is the point in condition (v). Put $\varphi(y) = Au(x_o,y) \wedge u(x_1,y)$. Then $\varphi \in S$ and $\varphi(x_1) \neq \varphi(x_2)$. Next put $\varphi(y) = u(x_o,y) \wedge 1$. Then $\varphi \in S$ and $\varphi(x) \neq \varphi(\partial)$ for any $x \in E$. Thus S separates the points of E_∂. Hence so does L. Let $f_1 \in L$, $f_2 \in L$, then $f_1 - f_2 = g_1 - g_2$ where $g_1 \in S$, $g_2 \in S$. Hence $g_1 \wedge g_2 \in S$, $g_1 + g_2 \in S$, $|f_1 - f_2| = |g_1 - g_2| = g_1 + g_2 - 2(g_1 \wedge g_2) \in L$, $f_1 \wedge f_2 = \frac{1}{2}\{f_1 + f_2 - |f_1 - f_2|\} \in L$, $f_1 \vee f_2 = f_1 + f_2 - (f_1 \wedge f_2) \in L$. Therefore L is a lattice. It is trivial that L is also a vector space.

Let K be a compact subset of E and let $L(K)$ denote the class of functions of L restricted to K. Let $\inf_{y \in K} u(x_o,y) = b > 0$. For any constant $c \geq 0$ put $\varphi(y) = \frac{c}{b}\, u(x_o,y) \wedge c$. Then $\varphi \in S$ and $\varphi = c$ on K. Therefore $L(K)$ contains all constants $c \geq 0$. It is also a lattice and a vector space. By a form of the Stone-Weierstrass theorem (see e.g. [5], p. 172), $L(K)$ is dense in $C(K)$ = the class of continuous functions on K.

Let $f \geq 0$ and have compact support $K \subset E$. For any $\varepsilon > 0$ there exists $g \in L$ such that $|f - g| \leq \varepsilon$ on K. Since $f \geq 0$, it is trivial that $|f - g^+| \leq \varepsilon$ on K. Since $g^+ \in C_o(E_\partial)$, there exists a compact K_1 such that $K \subset K_1 \subset E$ and $g^+ \leq \varepsilon$ on $E_\partial - K_1$. Hence, as before, there exists $h \in L$ such that $|f - h^+| \leq \varepsilon$ on K_1. Put $\varphi = g^+ \wedge h^+$; then $\varphi \in L$. We have $|f - \varphi| \leq \varepsilon$ on K; $|f - \varphi| = \varphi \leq h^+ \leq \varepsilon$ on $K_1 - K$; $|f - \varphi| = \varphi \leq g^+ \leq \varepsilon$ on $E_\partial - K_1$. Hence $|f - \varphi| \leq \varepsilon$ on E_∂. Since L is a vector space, it follows that L is dense in the class of continuous functions having compact support in E; hence it is also dense in $C_o(E_\partial)$.

Q.E.D.

4. The Dual Process

It is well known how to construct a Markov process on E_∂ with $(\hat{P}_t)$ as its transition semigroup. Let $\{Y_t, t \geq 0\}$ be such a process, and $G_t^o = \sigma(Y_s, 0 \leq s \leq t)$, $G^o = \bigvee_{0 \leq t < \infty} G_t^o$ its natural filtration. We can define, for each $y \in E_\partial$, the probability $\hat{P}^y$ and expectation $\hat{E}^y$ on G^o, associated with the process in the usual way. We proceed to show that there is a version of this process whose paths are almost surely (a.s.) right continuous in $[0,\infty)$ and have left limits in $(0,\infty]$. Although $(\hat{P}_t)$ is in general not a Feller semigroup, standard methods developed for the latter case in Chapter 2, §§2-4 of [3] can be adapted to the present situation with easy modifications. We shall indicate the main steps below.

(A) The process $\{Y_t\}$ is stochastically right continuous.

This follows from the weak (or just vague) right continuity of $\hat{P}_t$; see p. 50 of [3].

(B) Let R be the set of rational numbers in $[0,\infty)$. Then a.s. the restriction of the sample function $t \to Y_t$ to R has right limits in $[0,\infty)$ and left limits in $(0,\infty]$ everywhere.

Since $S \subset C_o(E_\partial)$ and the latter is a separably metrizable space, there is a countable subset D which is dense in S (with respect to the sup-norm of $C_o(E_\partial)$). If we use this set D instead of the $\mathbb{D}$ on p. 53 of [3], the same argument there works. Note that left limits exist at ∞, because a positive supermartingale has such a limit a.s.

(C) There is a version of (Y_t) whose sample functions are a.s. right continuous in $[0,\infty)$ and have left limits in $(0,\infty]$.

The argument is exactly the same as on p. 54 of [3]. From now on we shall use this version and refer to it as the dual process. Its lifetime will be denoted by $\zeta = T_{\{\partial\}}$. Then we have a.s.

$$Y_{t-} \in E \qquad \text{and} \qquad Y_t \in E \qquad \text{for} \quad 0 \leq t < \zeta .$$

The fact $\zeta > 0$ a.s. follows from $\hat{P}^y\{Y_o = y\} = \hat{P}_o(\{y\},y) = 1$ and right continuity. The rest is proved exactly as on pp. 54-55 of [3], if we use the member of S postulated in condition (v) or (v'), instead of the function $U^1\varphi$ there.

(D) For any coexcessive function f, the function $t \to f(X_t)$ is a.s. right continuous in $[0,\infty)$, and has left limits in $(0,\infty]$.

PROOF. Let g_n be as in (1) of §3, and let $K_j \uparrow E$, where each K_j is compact. Put

$$(1) \qquad \varphi_{nj}(y) = \int \xi(dx) g_n(x) 1_{K_j}(x)[u(x,y) \wedge ju(x_o,y) \wedge j] .$$

It follows from conditions (ii), (v) and (vii) that $\varphi_{nj} \in C_o(E_\partial)$. Hence $t \to \varphi_{nj}(Y_t)$ is an a.s. right continuous positive supermartingale. Since $g_n\hat{U} = \lim_{j \to \infty} \uparrow \varphi_{nj}$, the same is true of $t \to g_n\hat{U}(Y_t)$ by a theorem of Meyer's (see Theorem 5, §1.4 of [3], p. 32). Since $f = \lim_{n \to \infty} \uparrow g_n\hat{U}$, another application of the theorem establishes the right continuity of $t \to f(Y_t)$. The existence of left limits is then a consequence.

Meyer [4] proved that for a right continuous homogeneous Markov process, right continuity of all α-excessive functions along the paths is equivalent to the strong Markov property of the process. The proof given below follows his argument in one direction, but uses only 0-excessive functions.

(E) The dual process has the strong Markov property.

PROOF. Let T be optional with respect to $\{G^o_{t+}, t \geq 0\}$, and $T_n = 2^{-n}[2^n T + 1]$. Let f be positive continuous with compact support so that $f\hat{U} < \infty$ by (vii). We have by the simple Markov property, for each $y \in E$, $s \geq 0$:

$$\hat{E}^y\{f(Y_{T_n+s}) \mid G^o_{T_n}\} = f\hat{P}_s(Y_{T_n}) .$$

Integrating over s, we obtain

$$(2) \quad \hat{E}^y\{\int_{T_n+t}^{\infty} f(Y_s)ds \mid G^o_{T_n}\} = \int_t^{\infty} \hat{E}^y\{f(Y_{T_n+s}) \mid G^o_{T_n}\}ds = \int_t^{\infty} f\hat{P}_s(Y_{T_n})ds$$

$$= f\hat{U}\hat{P}_t(Y_{T_n}) .$$

Since $\hat{E}^y\{\int_0^{\infty} f(Y_s)ds\} = f\hat{U}(y) < \infty$, as $n \to \infty$ the first member in (2) converges to

$$\hat{E}^y\{\int_{T+t}^{\infty} f(Y_s)ds \mid G^o_{T+}\} = \int_t^{\infty} \hat{E}^y\{f(Y_{T+s}) \mid G^o_{T+}\}$$

by a well-known dominated convergence theorem for conditional expectations (see [6], Theorem 9.4.8). Since $f\hat{U}\hat{P}_t$ is coexcessive, the last member in (2) converges to $f\hat{U}\hat{P}_t(Y_T)$ as $n \to \infty$, by (D). The result is

$$(3) \quad \int_t^{\infty} \hat{E}^y\{f(Y_{T+s}) \mid G^o_{T+}\}ds = \int_t^{\infty} f\hat{P}_s(Y_T)ds .$$

The integrand in the left member of (3) is right continuous in s; so is that in the right member because $\hat{P}_s$ is weakly right continuous in $s \geq 0$. Hence it follows by differentiation of (3) that

$$\hat{E}^y\{f(Y_{T+t})|G^o_{T+}\} = f\hat{P}_t(Y_T) \tag{4}$$

for all $t \geq 0$. This implies the strong Markov property of the dual process.

(F) The dual process has the moderate Markov property.

PROOF. Let T be predictable with respect to $\{G^o_{t+}\}$, and let $\{T_n\}$ announce T. For each n, we have by (E), for any $f \in C_o(E_\partial)$:

$$\hat{E}^y\{f(Y_{T_n+t})|G^o_{T_n+}\} = f\hat{P}_t(Y_{T_n}) . \tag{5}$$

If $f \in S$, then $f\hat{P}_t$ is l.s.c. Letting $n \to \infty$ so that $Y_{T_n+t} \to Y_{T+t-}$ for both t and $t = 0$, we obtain

$$\hat{E}^y\{f(Y_{T+t-})|G^o_{T-}\} \geq f\hat{P}_t(Y_{T-}) \tag{6}$$

where $Y_{o-} = Y_o$. Now we let $t \downarrow\downarrow 0$, then $Y_{T+t-} \to Y_{T+} = Y_T$, while $f\hat{P}_t \to f$ by coexcessivity. It follows that

$$\hat{E}^y\{f(Y_T)|G^o_{T-}\} \geq f(Y_{T-}) . \tag{7}$$

Since $\{f(Y_t), G^o_{t+}, t \geq 0\}$ is a positive supermartingale, we have by the stopping theorem, for each n:

$$\hat{E}^y\{f(Y_T)|G^o_{T_n+}\} \leq f(Y_{T_n}) . \tag{8}$$

When $n \to \infty$ in (8), the result is an inequality which is reverse to (7). Therefore, (7) holds with equality for all $f \in S$, hence for all $f \in C_o(E_\partial)$ by Theorem 5. This implies $\hat{E}^y\{Y_T = Y_{T-}\} = 1$ by Lemma 1 on p. 66 of [3].

The quasi left continuity of the dual process follows by a general argument given on p. 70 of [3].

(G) Augmentation of $\{G_t^o, t \geq 0\}$.

Exactly as detailed on pp. 61-62 of [3], the natural filtration for the dual process can be augmented so that the new filtration $\{G_t, t \geq 0\}$ is right continuous. The strong and moderate Markov properties proved above are then valid for optional and predictable times with respect to the augmented filtration.

With these technical ramifications, we conclude that the dual process $\{Y_t, G_t; t \geq 0\}$ is a Hunt process with the semigroup $\{\hat{P}_t, t \geq 0\}$.

(H) Strong duality.

The duality relation in (4) of §2 implies the following. For each $\alpha \geq 0$, $f \in \mathcal{E}_+$, $g \in \mathcal{E}_+$, we have

$$\int (f\hat{U}^\alpha) g \, d\xi = \int f(U^\alpha g) d\xi ,$$

where

$$U^\alpha g = \int_0^\infty e^{-\alpha t} P_t g \, dt , \qquad f\hat{U}^\alpha = \int_0^\infty e^{-\alpha t} f\hat{P}_t \, dt .$$

Moreover, since

$$U(x,dy) = u(x,y)\xi(dy), \qquad \hat{U}(dx,y) = \xi(dx)u(x,y) ,$$

both $U(x,\cdot)$ and $\hat{U}(\cdot,y)$ are absolutely continuous with respect to ξ for each x and y; hence so are $U^\alpha(x,\cdot)$ and $\hat{U}^\alpha(\cdot,y)$, for $\alpha > 0$. Thus the hypotheses (referred to as those of "strong duality") in Theorem 1.4, §6.1 of [1] are satisfied. Therefore, all consequences of these hypotheses developed there apply to the two transient Hunt processes X and Y in this paper.

REFERENCES

1. R.M. BLUMENTHAL and R.K. GETOOR. *Markov Processes and Potential Theory*. Academic Press, New York, 1968.

2. K.L. CHUNG and K.M. RAO. A new setting for potential theory. *Ann. Inst. Fourier 30* (1980), 167-198.

3. K.L. CHUNG. *Lectures from Markov Processes to Brownian Motion.* Springer-Verlag, Berlin, 1982.

4. P.A. MEYER. *Processes de Markov*. Lecture Notes in Mathematics 26, Springer-Verlag, Berlin, 1967.

5. H.L. ROYDEN. *Real Analysis, 2nd Ed.* Macmillan Co., New York, 1968.

6. K.L. CHUNG. *A Course in Probability Theory, 2nd Ed.* Academic Press, New York, 1974.

K. L. Chung
Ming Liao
Department of Mathematics
Stanford University
Stanford, CA 94305

K. M. Rao
Department of Mathematics
University of Florida
302 Walker Hall
Gainesville, FL 32611

Seminar on Stochastic Processes, 1983
Birkhäuser, Boston, 1984

THEORIE GENERALE DU BALAYAGE

par

C. DELLACHERIE

On expose ici les parties les plus nouvelles des chapitres X et XI du 3e volume de "Probabilités et Potentiel", écrit conjointement avec P.A. Meyer et devant paraitre incessamment chez Hermann. Je me suis appliqué à expliquer les concepts, dégager les idées importantes mais je me suis contenté d'esquisser quelques démonstrations; j'espère qu'on me pardonnera de ne pas avoir développé ces dernières: d'abord, cela aurait fait double emploi avec celles du livre, et puis cela aurait été fort long.

Introduction

Soient $(E,\mathcal{E})$ un espace mesurable et $E^{\ddagger}$ l'ensemble des sous-probabilités m sur E (i.e. m est une mesure $\geqq 0$ de masse $\leqq 1$). Nous appellerons *maison de jeu* toute partie J de $E \times E^{\ddagger}$. Intuitivement, en supposant que, pour tout $x \in E$, les coupes J_x de J sont non vides et constituées de probabilités, E est l'espace des états x d'un joueur (la connaissance de l'état donnant tout renseignement utile sur le joueur - par exemple sa fortune f), et le joueur, dans l'état x, est autorisé à choisir un "jeu" dans J_x; une fois le jeu m_x

choisi dans J_x, le joueur se retrouve dans l'état $y \in E$ avec probabilité $m_x(dy)$ si bien qu'en moyenne sa fortune est alors $m_x(f) = \int f(y)\, m_x(dy)$. Nous reviendrons plus loin sur cette interprétation probabiliste et passons maintenant à la définition des notions fondamentales relatives à une maison de jeu: fonctions surmédianes, opérateur de réduction et (pré) ordre du balayage.

Nous associons d'abord à une maison de jeu J donnée un opérateur P, opérant sur les fonctions $\geqq 0$ sur E (sauf mention du contraire, le mot "fonction" désignera toujours une telle fonction par la suite) comme suit

$$Pf(x) = \sup_{m \in J_x} m^*(f) \quad (= 0 \ \textit{si} \ J_x = \emptyset)$$

où $m^*(f)$ est l'intégrale supérieure de f par rapport à m. Il est clair que P est un opérateur sous-linéaire, montant (i.e. $f_n \uparrow f \Rightarrow Pf_n \uparrow Pf$). *Une fonction* ($\geqq 0$) f *est dite surmédiane si on a* $Pf \leqq f$.

Voyons quelques exemples bien classiques:

1) Si J est le graphe d'un noyau sous-markovien de E dans E, alors P est ce noyau et les fonctions surmédianes mesurables sont les fonctions excessives par rapport à P.

2) Si $E = \mathbb{R}^n$ et si J_x est l'ensemble des probabilités uniformément réparties sur les boules (ou sphères) de centre x, alors les fonctions surmédianes mesurables sont les fonctions surharmoniques $\geqq 0$.

3) Si E est un convexe compact métrisable dans un e.l.c. et si J_x est l'ensemble des probabilités de barycentre x, les fonctions surmédianes sont les fonctions fortement concaves.

4) Si R est un préordre sur E (par exemple, une relation d'équivalence) et si $J = \{(x,\varepsilon_y): xRy\}$, alors les fonctions surmédianes

sont les fonctions décroissantes (et donc constantes sur les classes si R est une relation d'équivalence).

Revenons à notre maison de jeu J générale. L'ensemble $\mathfrak{S}$ des fonctions surmédianes est clairement un cône convexe contenant les constantes, stable pour les limites de suites croissantes et aussi pour les enveloppes inférieures de familles quelconques. De cette dernière propriété il résulte que *l'ensemble des fonctions surmédianes majorant une fonction* f *donnée admet un plus petit élément appelé la réduite de* f *et noté* Rf. On vérifie sans peine qu'on peut calculer la réduite de f à l'aide de l'opérateur P comme suit

THEOREME 1. *Soit* Q *l'opérateur défini par* $Qf = f \vee Pf$ *pour toute fonction* f. *On a alors, pour toute fonction* f,

$$Rf = \lim_n \uparrow Q^n f$$

si bien que l'opérateur de réduction R *est sous-linéaire, montant, supérieur à l'identité, et idempotent.*

Et dans ce cadre général, on a les propriétés suivantes (classiques dans le cas de l'ex. 1))

THEOREME 2. a) *On a* $Rf = f \vee PRf$.

b) *Si* Rf *est finie partout, alors, pour tout* $t < 1$, *on a*

$$Rf = R(f1_A) \quad \text{où} \quad A = \{tRf < f\}.$$

Ceci dit, la théorie générale du balayage est, en gros, l'étude du cône $\mathfrak{S}$ et de l'ordre du balayage associé qui, dans la cadre trop général dans lequel nous nous sommes placés (nous n'avons fait aucune hypothèse de mesurabilité sur J), peut se définir ainsi: pour $\lambda,\mu \in E^{\ddagger}$, on dit que μ *est une balayée de* λ *et on note* $\lambda \dashv \mu$ *ssi*

$$\forall f \in \mathcal{S} \quad \lambda_*(f) \geqq \mu_*(f)$$

où ici nous utilisons l'intégrale inférieure. En fait cette dernière expression équivaut à $\lambda_*(Rg) \geqq \mu(g)$ pour toute g $\mathcal{E}$-mesurable et bientôt nous aurons assez de fonctions surmédianes mesurables pour faire disparaitre cette intégrale inférieure. Nous garderons par contre l'intégrale supérieure introduite plus haut car il est commode d'avoir Pf et Rf définies pour toute fonction f.

Chacun devine qu'on n'ira pas loin si on n'a pas à sa disposition une classe suffisamment riche de fonctions mesurables stable pour les opérateurs P et R. Mais la situation n'est pas simple: dans l'exemple 4), avec $E = [0,1]$ et une relation d'équivalence à graphe compact, $R1_B$ est, pour toute partie B de E, l'indicatrice du saturé A de B pour la relation; pour B borélien, A n'est pas forcément borélien (il est cependant analytique et donc universellement mesurable) et, pour B universellement mesurable, A n'est pas forcément universellement mesurable. On va s'en sortir grâce à la notion, absolument indispensable ici, de fonction analytique i.e. de fonction f telle que, pour tout $t \geqq 0$, l'ensemble $\{f \geqq t\}$ soit analytique (nous verrons ci-dessous une définition équivalente mieux adaptée à notre situation).

Hypothèses de Régularité

Désormais notre espace d'états $(E,\mathcal{E})$ sera un espace métrisable compact muni de sa tribu borélienne. En fait, comme tout espace polonais (ou plus généralement lusinien) est boréliennement isomorphe à un espace métrisable compact, la plupart des résultats que nous donnerons s'étendent à ces espaces (et même aux espaces sousliniens avec guère plus de travail).

Rappelons qu'une partie A de E est dite *analytique* s'il existe un espace compact métrisable F et un borélien B de $E \times F$ tel que $A = \pi_E(B)$. De même, une fonction f sur E est dite *analytique* s'il existe un espace compact métrisable F et une fonction borélienne g sur $E \times F$ telle que $f(x) = \sup_{y \in F} g(x,y)$ pour tout $x \in E$ (et, du coup, si g est seulement supposée analytique sur $E \times F$, on obtient sans peine que f est encore analytique sur E). Toute fonction analytique est universellement mesurable, et l'ensemble des fonctions analytiques est un cône convexe contenant les fonctions boréliennes, stable pour la multiplication (rappelons que nos fonctions sont $\geqq 0$) et pour les sup et inf dénombrables (opérations "interdites": soustraction et division; pour f analytique $\leqq 1$, $1 - f$ et $1/f$ ne sont analytiques que si f est borélienne).

Ceci fait, nous munissons $E^{\ddagger}$ de la topologie vague, qui en fait un espace métrisable compact. Dans ces conditions, il est clair que si J est une maison de jeu borélienne et f une fonction borélienne alors Pf est une fonction analytique; plus généralement, on vérifie aisément que, si J est analytique et f analytique, alors Pf, Qf et Rf sont analytiques. Ainsi, si J est analytique, le cône $\mathcal{S}_a$ des fonctions surmédianes analytiques est conséquent et on a

$$\lambda \dashv \mu \quad ssi \quad \forall f \in \mathcal{S}_a \quad \lambda(f) \geqq \mu(f)$$

en effet, supposons la propriété de droite vérifiée, et soient $h \in \mathcal{S}$ et g borélienne $\leqq h$ telle que $\lambda_*(h) = \lambda(g)$, $\mu_*(h) = \mu(g)$. On a $g \leqq Rg \leqq h$ et $Rg \in \mathcal{S}_a$ et donc $\lambda_*(h) = \lambda(Rg) \geqq \mu(Rg) = \mu_*(h)$, d'où la conclusion. Nous verrons plus loin que, pour définir l'ordre du balayage, il suffit en fait de considérer le cône $\mathcal{S}_b$ des fonctions surmédianes boréliennes: c'est loin d'être évident car, en dehors du cas où J est une réunion dénombrable de compacts (où l'on sait que Rf est borélienne pour f s.c.s.), on ne voit pas a priori comment construire des éléments de $\mathcal{S}_b$ non triviaux.

Par commodité, il nous arrivera encore par la suite de considérer des maisons de jeux quelconques (relatives à notre bon espace d'états), mais les résultats porteront sur les maisons analytiques. Ainsi nos hypothèses de régularité sont: E métrisable compact et J analytique dans $E \times E^{\ddagger}$.

Interprétation Probabiliste

Soit donc J une maison de jeu analytique. Nous supposons ici que chaque coupe J_x est constituée de probabilités et contient la mesure de Dirac ε_x (on dit alors que la maison est *quittable*: le joueur peut rester autant de temps qu'il veut dans son état; du point de vue probabiliste, cela veut dire qu'on pourra faire opérer les temps d'arrêt). On appelle ***stratégie*** toute suite $s = (s_n)$ de noyaux $s_n: E^n \to E$ vérifiant la condition

$$\forall \vec{x} = (x_1, \dots, x_n) \in E^n \quad s_n(\vec{x}, dy) \in J_{x_n}$$

la mesure $s_1(x,dy)$ représente le jeu choisi par le joueur à l'instant 1 quand il est dans l'état initial x; la mesure $s_2(x_1,x_2,dy)$

représente le jeu choisi par le joueur à l'instant 2 si le résultat du premier jeu l'a mis dans l'état x_2, etc. Autrement dit une stratégie est un programme de jeu que le joueur a choisi à l'avance en prévoyant tous les cas possibles.

La stratégie s étant donnée, on construit comme d'habitude, pour toute loi initiale λ sur E, une unique probabilité $\mathbb{P}_s^\lambda$ sur $\Omega = E^{\mathbb{N}}$ telle que

$$\mathbb{P}_s^\lambda\{X_0 \in B\} = \lambda(B) \quad \textit{pour} \quad B \in \mathcal{E}$$

$$\mathbb{P}_s^\lambda\{X_{n+1} \in B | \mathcal{F}_n\} = s_{n+1}(X_0, \ldots, X_n, B)$$

où (X_n) désigne la suite des coordonnées et $(\mathcal{F}_n)$ sa filtration. Il est clair qu'une fonction universellement mesurable g est surmédiane pour J ssi, pour toute loi initiale λ et toute stratégie s, le processus $(g(X_n))$ est une surmartingale (généralisée) par rapport à $\mathbb{P}_s^\lambda$. Cela implique que, pour toute loi initiale λ, toute stratégie s, et tout temps d'arrêt T, la loi μ de $X_T 1_{\{T<\infty\}}$ relativement à $\mathbb{P}_s^\lambda$ est une balayée de λ.

Soit maintenant f une fonction universellement mesurable que nous interpréterons comme la fortune de joueur et posons, pour toute loi initiale λ,

$$\overline{f}(\lambda) = \sup_s \sup_T \mathbb{E}_s^\lambda[f(X_T) 1_{\{T<\infty\}}]$$

où s parcourt l'ensemble des stratégies et T celui des temps d'arrêt (comme on a un jeu quittable, on peut en fait se contenter des temps d'arrêt constants et des stratégies de longueur finie). La quantité

$\overline{f}(x) = \overline{f}(\varepsilon_x)$ est le mieux que puisse espérer avoir, en moyenne, le joueur quand il quitte la maison de jeu après y être entré dans l'état x. Et le théorème de section des ensembles analytiques permet de démontrer le résultat suivant dû à Strauch.

THEOREME 3. *Si* f *est analytique, on a pour toute loi* λ

$$\overline{f}(\lambda) = \lambda(Rf)$$

et, en particulier, sur E *on a* $\overline{f} = Rf$

COROLLAIRE. *Pour toute loi* λ, *on a, pour* f *analytique*,

$$\lambda(Rf) = \sup_{\lambda \dashv \mu} \mu(f).$$

Ce dernier résultat, purement analytique, est vrai sans restriction sur les masses des mesures de la maison de jeu analytique J.

Il est naturel de chercher, dans ce contexte, s'il existe des stratégies optimales ou tout au moins ε-optimales d'un type particulier. Voici le meilleur résultat général que l'on connaisse dans cette direction, dû à Sudderth et Ornstein; il assure l'existence d'une stratégie ε-optimale s qui soit markovienne (i.e. $s_n(\vec{x},dy)$ ne dépend que de x_n) et stationnaire (i.e. s_n ne dépend pas de n).

THEOREME 4. *Supposons* f *analytique et* Rf *finie partout. Alors pour toute loi* λ *et tout* $\varepsilon > 0$ *il existe un noyau markovien* N *de* E *dans* E, *de graphe contenu dans* J, *tel qu'on ait*

$$R_N f \geqq (1-\varepsilon)Rf \qquad \lambda\text{-p.p.}$$

où R_N *est l'opérateur de réduction associé au noyau* N.

Il s'agit là (sous une forme légèrement améliorée) d'un résultat assez ancien, de démonstration restant difficile, et dont on ne connait aucune application...

Opérateurs Capacitaires en Théorie du Balayage

Les outils essentiels dans l'étude du balayage sont le théorème de Hahn-Banach et le théorème de capacitabilité. Cela n'est pas nouveau: c'était déjà le cas dans la théorie bien connue du balayage par rapport à un cône de fonctions continues (sur laquelle nous reviendrons plus loin). Cependant, si nous n'avons rien à ajouter du côté de Hahn-Banach (du moins, pour le moment: plus loin, nous ferons usage d'un théorème de Mokobodzki qui, par certains côtés, est une extension du théorème de Hahn-Banach), nous aurons par contre à utiliser dans notre étude les développements les plus récents de la théorie des capacités. Aussi, sans chercher à être exhaustif, nous allons tenter ici d'expliquer de quoi il s'agit en restant le plus près possible de nos maisons de jeux. Deux mots encore sur les notations avant de s'y mettre. Pour notre étude, il sera nécessaire de considérer nos opérateurs P, Q, R comme portant sur deux arguments: la maison J et la fonction f. D'où les notations P_J ou $P(J,\cdot)$, etc, quand il y aura doute sur J (et de même nous parlerons de l'ensemble $\mathcal{S}(J)$ des fonctions J-surmédianes).

Si J est une maison de jeu compacte, l'opérateur P_J a les propriétés suivantes, qu'on résume en disant que P_J est *capacitaire*

i) il est croissant:

$$f_1 \leqq f_2 \Rightarrow P_J f_1 \leqq P_J f_2$$

2) il est montant:

$$f_n \uparrow f \Rightarrow P_J f_n \uparrow P_J f$$

3) il est descendant sur l'ensemble des fonctions s.c.s. (qui est l'analogue fonctionnel de l'ensemble des compacts de E):

$$g_n \downarrow g \Rightarrow P_J g_n \downarrow P_J g$$

si les g_n sont s.c.s.

4) enfin, $P_J g$ est s.c.s. si g est s.c.s.

Les propriétés 3) et 4) ne sont pas tout à fait évidentes; elles résultent cependant aisément du fait que, pour g s.c.s., la fonction $m \to m(g)$ sur $E^{\ddagger}$ est s.c.s., et du lemme de Dini-Cartan (l'analogue pour les fonctions s.c.s. de la propriété d'intersection finie des compacts). Les propriétés 1), 2), 3) sont ponctuelles, et expriment que, pour tout $x \in E$, la fonctionnelle $f \to P_J f(x)$ est une capacité de Choquet (à ceci près que, d'habitude, l'argument d'une capacité est un ensemble plutôt qu'une fonction; il s'agit ici d'une extension du concept analogue à celle faisant passer de la mesure à l'intégrale). Ainsi, notre opérateur P_J est aux capacités ce qu'un noyau est aux mesures, et la propriété de régularité 4) correspond en quelque sorte à la propriété de Feller en théorie de la mesure.

De même, pour J compact, l'opérateur Q_J est capacitaire (Q_J est d'ailleurs égal à $P_{J'}$, où J' est la maison de jeu quittable engendrée par J i.e. $J' = J \cup \{(x,\varepsilon_x): x \in E\}$), ainsi que ses puissances Q_J^n (le composé de deux opérateurs capacitaires étant encore de toute évidence capacitaire). Par contre, l'opérateur limite R_J ne l'est pas forcément (perte des propriétés 3) et 4)). Voyons cela de plus près. Désignons, pour J quelconque, par $\tilde{J}$ la maison *saturée* de J définie par $(x,m) \in \tilde{J} \Leftrightarrow \varepsilon_x \dashv_J m$; c'est la plus grande maison admettant les mêmes fonctions universellement mesurables surmédianes que J, et nous verrons plus loin qu'elle est analytique si J l'est (c'est loin d'être évident, même pour J compacte). D'après le corollaire du théorème 3, si la maison J est analytique, on a

$$R_J f = P_{\tilde{J}} f = R_{\tilde{J}} f$$

pour toute fonction analytique f, et comme ce sont les seules fonctions qui nous intéressent, nous écrirons abusivement $R_J = P_{\tilde{J}} = R_{\tilde{J}}$. De l'égalité précédente, il résulte que, pour J analytique, R_J est capacitaire si $\tilde{J}$ est compacte; la réciproque est conséquence immédiate du résultat suivant, qui n'est pas vraiment nouveau mais qui illustre bien, dans un cas simple, l'utilisation combinée du théorème de Hahn-Banach et du théorème de capacitabilité.

THEOREME 5. *Soit* S *un opérateur capacitaire, sous-linéaire, tel que* $S1 \leq 1$. *On a alors* $S = P_J$ *où* J *est la maison de jeu, compacte, définie par*

$$(x,m) \in J \quad \textit{ssi} \quad \forall f \in \mathbf{A} \quad Sf(x) \geq m(f)$$

où $\mathbf{A}$ *est l'ensemble des fonctions analytiques. De plus, si* S *est idempotent et supérieur à l'identité, alors* J *est saturée et* $S = R_J$.

DEMONSTRATION. Le théorème de capacitabilité pour les opérateurs capacitaires nous assure que, pour f analytique, Sf est encore analytique et que

$$Sf = \sup\{Sg \colon g \leq f,\ g \text{ s.c.s.}\}$$

Il en résulte que, dans la définition de J, on peut se contenter de prendre les f s.c.s.; mais, S descendant sur las fonctions s.c.s., on peut même se contenter d'y prendre les f continues. Il est alors clair, d'une part, que la maison J est compacte, et d'autre part, grâce au théorème de Hahn-Banach, qu'on a $Sf = P_J f$ pour f continue. Mais, S et P_J étant capacitaires, le théorème de capacitabilité joint à la descente sur les fonctions s.c.s. nous donne $Sf = P_J f$ pour f analytique, d'où l'égalité (abusive) de l'énoncé. La deuxième partie de l'énoncé est triviale.

On rencontre de manière naturelle des opérateurs de réduction capacitaires dans la théorie du balayage par rapport à un cône de fonctions continues, théorie dont nous rappelons maintenant, briévement, les premiers pas. Soit $\mathfrak{C}$ un cône convexe de fonctions continues contenant la constante 1 (nos fonctions sont $\geqq 0$ mais on peut toujours se ramener à ce cas). On associe à $\mathfrak{C}$ une maison J par

$$(x,m) \in J \quad \textit{ssi} \quad \forall f \in \mathfrak{C} \qquad f(x) \geqq m(f)$$

La maison J est compacte, saturée (toute $f \in \mathfrak{C}$ étant surmédiane) et donc $R_J (= P_J)$ est capacitaire. Comme au théorème 5, un argument de capacitabilité montre alors qu'on a

$$(°) \qquad \lambda \dashv \mu \quad \textit{ssi} \quad \forall g \ \textit{s.c.s.}\ \lambda(Rg) \geqq \mu(Rg).$$

Par ailleurs, une application du théorème de Hahn-Banach montre que

$$(°°) \qquad Rg = \inf\{f \in \mathfrak{C} : f \geqq g\}$$

pour toute fonction g s.c.s. (c'est ici qu'intervient le fait que $\mathfrak{C}$ soit un cône convexe). On déduit en particulier de (°) et (°°) que $\mathfrak{C}$ définit le balayage (i.e. $\lambda \dashv \mu$ ssi $\forall f \in \mathfrak{C}\ \lambda(f) \geqq \mu(f)$) s'il est infstable - et stabiliser $\mathfrak{C}$ pour les inf. finis ne change pas J.

Ceci dit, un aménagement, de la démonstration du théorème de séparation d'Urysohn permet d'établir que, réciproquement, tout opérateur de réduction capacitaire est associé à un tel cône (il s'agit là d'un résultat nouveau) si bien que l'on peut énoncer

THEOREME 6. *Soit* J *une maison de jeu analytique. L'opérateur* R_J *est capacitaire ssi le cône* $\mathcal{S}_c(J)$ *des fonctions continues* J-*sur-médianes définit l'ordre du balayage, i.e. on a*

$$\lambda \dashv_J \mu \quad ssi \quad \forall f \in \mathcal{S}_c(J) \quad \lambda(f) \geqq \mu(f)$$

La condition nécessaire, qui assure l'existence de nombreuses fonctions surmédianes continues, n'est évidemment intéressante que si l'on connait des moyens autres que celui exposé ci-dessus pour construire des opérateurs de réduction capacitaires. Nous en verrons un un peu plus loin.

Nous poursuivons l'étude de nos opérateurs en y faisant maintenant varier l'argument J, et, à cette occasion, allons rencontrer la notion importante d'opérateur capacitaire à deux arguments (qui n'a pas d'analogue intéressant en théorie de la mesure, toute bimesure se ramenant naturellement à une mesure sur un produit).

THEOREME 7. *L'opérateur* $(J,f) \to P(J,f)$ *est un opérateur capacitaire à deux arguments, i.e. a les propriétés suivantes:*

a) *il est croissant en les deux arguments*

b) *il est montant en les deux arguments*

c) *il est descendant en les deux arguments quand ceux-ci sont resp. compact et s.c.s.*

d) *enfin,* P(K,g) *est s.c.s. quand* K *est compact et* g *s.c.s..*

Les propriétés a), b) sont évidentes; le seule nouveauté en c), d) par rapport aux propriétés 3), 4) vues précédemment est la descente en le premier argument compact quand le second est s.c.s.: c'est encore une conséquence du lemme de Dini-Cartan. Notons que, pour J analytique fixé, l'opérateur $P_J(\cdot)$ à un argument est obtenu à partir de l'opérateur capacitaire $P(\cdot,\cdot)$ à deux arguments en y fixant un argument analytique: on résume cela en disant que P_J est un *opérateur analytique* (de manière générale est analytique un opérateur à m arguments obtenu en fixant $n-m$ arguments analytiques dans un opérateur capacitaire à n arguments). Bien entendu, les opérateurs $(J,f) \to Q^n(J,f)$ sont aussi capacitaires pour tout $n \in \mathbb{N}$, mais leur limite croissante $(J,f) \to R(J,f)$ ne l'est jamais (perte des propriétés c), d)) si E est infini. Ce dernier opérateur, que nous étudierons plus précisément un peu plus loin, est cependant toujours un opérateur analytique (pour le voir, il faut introduire un opérateur capacitaire à trois arguments, le troisième étant pris dans l'espace métrisable compact $\overline{\mathbb{N}}$ puis fixé égal à $\mathbb{N}$), ainsi donc que l'opérateur R_J pour J fixé analytique. Ceci dit, les opérateurs capacitaires (et plus généralement, de manière évidente, les opérateurs analytiques) vérifient les deux propriétés importantes suivantes que nous écrivons pour un opérateur S à deux arguments J,f pour fixer les idées

A) *Théorème de capacitabilité*

Si les arguments J,f sont analytiques, alors $S(J,f)$ est analytique et on a l'approximation par en dessous

$$S(J,f) = \sup\{S(K,g):\ J \supseteq K \textit{ compact},\ f \geqq g \textit{ s.c.s.}\}$$

B) *Théorème de séparation*

Si les arguments J,f sont analytiques et si h est une fonction borélienne (ou plus généralement coanalytique i.e. $1/h$ est analytique) majorant $S(J,f)$, alors il existe un borélien I contenant J et une fonction borélienne e majorant f tels que h majore encore $S(I,e)$.

On voit très bien à quoi peut servir le théorème de capacitabilité; nous l'avons d'ailleurs déjà utilisé dans la démonstration du théorème 5. Par contre, on voit moins bien à quoi peut servir ce bizarre théorème de séparation; en fait, il fournit en particulier un moyen puissant pour parler des fonctions boréliennes surmédianes. Voici un premier exemple, qui nous dit que toute maison de jeu analytique possède beaucoup de fonctions surmédianes boréliennes (sans nous donner, cependant, un moyen commode pour en construire!).

THEOREME 8. *Soient* J *une maison de jeu analytique,* f *une fonction* J*-surmédiane analytique et* h *une fonction borélienne majorant* f. *Il existe alors une fonction* J*-surmédiane borélienne* g *telle que l'on ait* $f \leqq g \leqq h$.

DEMONSTRATION. L'opérateur $R = R_J$ est analytique. Donc, d'après le théorème de séparation, il existe une fonction borélienne $h_1 \geqq f$ telle que $Rh_1 \leqq h$ si bien qu'on a $f \leqq h_1 \leqq Rh_1 \leqq h$. On recommence avec h_1 à la place de h, etc; par récurrence, on construit ainsi une suite (h_n) de fonctions boréliennes telle qu'on ait

$$f \leqq h_{n+1} \leqq Rh_{n+1} \leqq h_n \leqq Rh_n \leqq h$$

Soit alors $g = \inf_n h_n = \inf_n Rh_n$; g est borélienne, coincée entre f et h, et on a $g \leqq Rg \leqq g$ d'où $Rg = g$ si bien que g est surmédiane.

COROLLAIRE. *Si* J *est analytique, le cône* $\mathcal{S}_b(J)$ *des fonctions surmédianes boréliennes définit l'ordre du balayage, i.e. on a*

$$\lambda \dashv \mu \quad \textit{ssi} \quad \forall g \in \mathcal{S}_b \ \lambda(g) \geqq \mu(g).$$

DEMONSTRATION. Nous avons vu que la relation $\dashv$ est definie par le cône $\mathcal{S}_a$; il suffit donc de prouver que tout élément f de $\mathcal{S}_a$ est $\lambda + \mu$-p.p. égal à un élément g de $\mathcal{S}_b$. Et cela résulte immédiatement du théorème (prendre pour h une fonction borélienne $\geqq f$ partout et égale à f $(\lambda+\mu)$-p.p.; cela existe car f est universellement mesurable).

Pour poursuivre l'étude de la relation de balayage (un de nos objectifs est de prouver que la saturée d'une maison analytique est encore analytique), il nous est nécessaire de savoir approcher les opérateurs de réduction analytiques par des opérateurs de réduction capacitaires. C'est ce que nous allons faire maintenant.

Pour J fixée quelconque et $\theta \in [0,1[$, nous désignerons par θJ la maison $\{(x,\theta m): (x,m) \in J\}$. Une fonction f est donc θJ-surmédiane ssi on a $\theta P_J f \leqq f$; on reconnait donc là une notion familière dans le cadre de l'exemple 1) vu plus haut. Il est clair qu'une fonction g est J-surmédiane ssi elle est θJ-surmédiane pour tout $\theta < 1$ et, plus généralement, on voit sans peine que $R_J f = \lim_{\theta \to 1} \uparrow R_{\theta J} f$ pour toute fonction f. Le lemme suivant, très simple, dévoile l'intérêt de l'adjonction de ce paramètre θ .

LEMME. *Pour toute maison* J *et tout* $\theta \in [0,1[$ *on a, pour toute fonction* f *de norme uniforme* M,

$$R_{\theta J}f - Q^n_{\theta J}f \leqq M\theta^n/(1-\theta)$$

pour tout $n \in \mathbb{N}$.

DEMONSTRATION. Ayant fixé J et θ, nous notons P, Q, R les opérateurs associés à la maison θJ. On a bien évidemment

$$Qf - f = (Pf-f)^+ \leqq Pf \leqq \theta M$$

et, de manière générale, en utilisant le fait que P est majoré par Q et est sous-linéaire,

$$Q^{n+1}f - Q^n f = (PQ^n f - Q^n f)^+ \leqq PQ^n f - PQ^{n-1}f \leqq P(Q^n f - Q^{n-1}f)\,.$$

On en tire par récurrence

$$Q^{n+1}f - Q^n f \leqq \theta^{n+1}M$$

d'où l'énoncé.

Cette approximation de R par les Q^n uniformément en J et f pour θ fixé entraine aisément le théorème-clé suivant.

THEOREME 9. *Pour tout* $\theta \in [0,1[$, *l'opérateur* $(J,f) \to R(\theta J,f)$ *est un opérateur capacitaire.*

Remarquons au passage que ceci implique que, pour J compact et $\theta < 1$, $R_{\theta J}$ est capacitaire. Le théorème 6 entraine alors qu'il y a beaucoup de fonctions continues θJ-surmédianes; mais, bien entendu, il peut n'en rester aucune pour $\theta = 1$.

COROLLAIRE. *Si* J *est analytique, il existe* $\check{J}$ *analytique comprise entre* J *et sa saturée telle qu'on ait* $R_J = P_{\check{J}} = R_{\check{J}}$

Ce résultat, qui est une étape de la démonstration de l'analyticité de la saturée de J, n'est pas, malgré le nom donné, une conséquence immédiate du théorème et de ce qui le précède: il faut, pour l'obtenir, appliquer à l'opérateur capacitaire $(J,f) \to R(\theta J,f)$ une forme du théorème de capacitabilité plus sophistiquée que celle que nous avons donnée, à savoir "les opérateurs capacitaires traversent les schémas de Souslin privilégiés". Les initiés comprendront; les autres auront envie d'apprendre... .

Balayage Séparable

Nous allons élucider ici la structure du balayage dans un cas très particulier mais néanmoins digne d'intérêt. La maison de jeu J est dite (à balayage) *séparable* s'il existe une suite (f_n) de fonctions boréliennes telle qu'on ait

$$\lambda \dashv_J \mu \quad ssi \quad \forall n \ \lambda(f_n) \geqq \mu(f_n) .$$

Il est clair que les f_n sont alors surmédianes et que la relation de balayage est borélienne (ainsi donc que la saturée de J). D'autre part, quitte à transformer de manière évidente les f_n, on peut supposer que l'ensemble des f_n contient la constante 1, est stable pour les inf. finis, et est constitué de fonctions bornées: nous dirons alors, pour abréger, que (f_n) est une *bonne* suite.

Voici deux exemples familiers:

i) Si J est la maison de jeu associée à un noyau sous-markovien

P de E dans E, le balayage est séparable dès que l'opérateur potentiel G de P est propre. En effet, on sait alors qu'il existe une fonction borélienne $h > 0$ partout telle que $Gh < \infty$ partout et qu'on a $\lambda \dashv \mu$ ssi $\lambda G \geqq \mu G$. Il suffit donc de prendre pour (f_n) l'image par $G(\cdot/h)$ d'une bonne suite de fonctions boréliennes engendrant $\mathcal{E}$.

ii) La maison J est séparable si son opérateur de réduction R est capacitaire. En effet, d'après le théorème de capacitabilité et la descente de R sur les fonctions s.c.s. (cf la démonstration du th. 5) on a $\lambda \dashv \mu$ ssi $\lambda(Rf) \geqq \mu(Rf)$ pout toute f continue. Il suffit donc de prendre pour (f_n) l'image par R d'une bonne suite de fonctions continues dense pour la convergence uniforme dans les fonctions continues. Bien entendu, on aurait pu procéder plus rapidement en utilisant le théorème 6 et en prenant une bonne suite dense dans l'ensemble des fonctions surmédianes continues.

En dehors de ces exemples, il semble difficile de vérifier qu'un balayage donné a priori est séparable. Pourtant, le résultat suivant montre qu'on peut construire beaucoup de maisons séparables.

THEOREME 10. *Soient* (f_n) *une suite de fonctions boréliennes et* J *la maison de jeu saturée définie par*

$$(x,m) \in J \quad ssi \quad \forall n \ f_n(x) \geqq m(f_n).$$

Alors la maison J *est séparable et, si* (f_n) *est une bonne suite, on a*

$$\lambda \dashv \mu \quad ssi \quad \forall n \ \lambda(f_n) \geqq \mu(f_n).$$

DEMONSTRATION. On peut évidemment supposer que (f_n) est une bonne suite sans changer la maison J. On sait que c'est vrai si les f_n sont continues (considérer le cône $\mathfrak{C}$ engendré par les (f_n)), et la seule méthode de démonstration que nous connaissions est de se ramener à ce cas par compactification. Les f_n étant boréliennes, bornées, on sait qu'on peut considérer E comme partie borélienne d'une espace métrisable compact $\hat{E}$ qui induit sur E la même structure borélienne (mais une topologie plus fine) que l'initiale, et tel que chaque f_n admette un unique prolongement $\hat{f}_n$ en une fonction continue. Soit alors $\hat{\mathfrak{C}}$ le cône engendré par les $\hat{f}_n$. On vérifie sans peine que la maison J est la trace sur $E \times E^{\ddagger}$ de la maison $\hat{J}$ sur $\hat{E} \times \hat{E}^{\ddagger}$ associée à $\hat{\mathfrak{C}}$ si bien que $\mathfrak{S}$ est exactement l'ensemble des restrictions à E des éléments de $\hat{\mathfrak{S}}$, d'où la relation $\dashv_J$ est aussi la trace sur E de la relation $\dashv_{\hat{J}}$. Comme cette dernière est définie par $(\hat{f}_n)$, on peut conclure.

Par la même méthode de compactification, on déduit d'une conséquence classique du théorème de Strassen le résultat suivant .

THEOREME 11. *Soit* J *une maison séparable. On a* $\lambda \dashv_J \mu$ *ssi il existe un noyau sous-markovien* N *de* E *dans* E *tel que* $\mu = \lambda N$ *et qu'on ait* $\varepsilon_x \dashv_J \varepsilon_x N$ *pout tout* $x \in E$.

Nous terminons en montrant que la séparabilité d'une maison J entraine la séparabilité du cône $\mathfrak{S}_b$ (au sens où on l'entend pour une tribu). Le lecteur pourra à partir de cela retrouver le théorème de Blackwell pour les sous-tribus séparables de $\mathbf{E}$ en partant d'une sous-algèbre dénombrable $\mathbf{B}$ de $\mathbf{E}$ et en considérant la maison J définie par les indicatrices des éléments de $\mathbf{B}$ (les fonctions surmédianes sont alors celles qui sont constantes sur chaque atome de la tribu engendrée par $\mathbf{B}$).

THEOREME 12. *Soit* (f_n) *une bonne suite de fonctions boréliennes définissant la maison de jeu séparable et saturée* J. *Alors le cône* $\mathcal{S}_b$ *des fonctions surmédianes est le plus petit cône contenant les* f_n *et stable pour les limites de suites croissantes ou décroissantes.*

DEMONSTRATION. Nous désignerons par $\mathcal{S}^\circ$ le plus petit cône contenant les f_n et stable pour les limites monotones: il est clair que $\mathcal{S}^\circ$ est contenu dans $\mathcal{S}_b$. Pour la réciproque, on commence par supposer que les f_n sont continues (même dans ce cas, le résultat est nouveau quoique inspiré des travaux de Preiss sur la génération des convexes boréliens dans $\mathbb{R}^n$). Alors l'opérateur de réduction est capacitaire et une application du théorème de séparation, ou plutôt une adaptation de sa démonstration utilisant à la fois que R est capacitaire et idempotent (nous ne donnerons pas de détails ici), permet d'obtenir le résultat de séparation suivant: si f est analytique et h borélienne $\geqq$ Rf, alors il existe $g \in \mathcal{S}^\circ$ telle que $h \geqq g \geqq Rf$. Mais, si on prend $f \in \mathcal{S}_b$ et $h = f$, on obtient $f = g$, d'où $f \in \mathcal{S}^\circ$. Pour terminer, on considère une compactification comme ci-dessus. On a là encore à utiliser un argument de séparation (fourni cette fois par l'énoncé B) plus haut) pour être assuré que tout élément de $\mathcal{S}_b$ est trace sur E d'un élément de $\hat{\mathcal{S}}_b$.

Ajoutons une remarque: ce qui vient d'être fait dans le cadre E métrisable compact (ou plus généralement souslinien) s'étend "à quelques $\mathbf{P}$-p.p. près" au cas où $(E,\mathbf{E})$ est un espace mesurable radonien (i.e. mesurablement isomorphe à une partie universellement mesurable d'un espace métrisable compact). Cela vient du fait que, pour toute mesure

$\mathbb{P}$ et toute suite (f_n) de fonctions universellement mesurables sur un tel espace $(E,\mathcal{E})$, il existe $F \in \mathcal{E}$ portant $\mathbb{P}$ tel que $(F,\mathcal{E}_{|F})$ soit isomorphe mesurablement à un espace métrisable compact et que chaque f_n restreinte à F soit $\mathcal{E}_{|F}$-mesurable.

Retour au Cas Général

Nous revenons au cas où J est une maison analytique quelconque. Nous allons montrer que la relation de balayage est analytique et qu'on a encore, sous une forme affaiblie, le "théorème de Strassen" (cf l'énoncé du th. 11).

Nous commençons par une longue digression, nécessaire. Soit H une partie de $E^{\ddagger}$ et définissons une forme sous-linéaire p_H en posant pour toute fonction f

$$p_H(f) = \sup_{m\in H} m^*(f).$$

On reconnait là, du moins pour H et f analytiques, les formes que l'on rencontre en évaluant nos opérateurs P, Q, R (pour R, cf le cor. du th. 9) en un point, et comme nous ne nous intéressons qu'aux fonctions analytiques, nous identifierons, comme plus haut pour nos opérateurs, deux telles formes si elles coincident sur les fonctions analytiques. Par approximation par en dessous (triviale ici car il s'agit de sup. de mesures), on voit que p_H est bien déterminée, parmi les formes sous-linéaires du même type, par sa restriction aux fonctions boréliennes et même s.c.s., mais on ne peut pas aller jusqu'aux fonctions continues (considérer le cas où H n'est formée que de mesures de Dirac) sans hypothèse supplémentaire (par exemple, H compact, qui implique que p_H descend sur les fonctions s.c.s.). Voyons cela de plus près; notons p_H^C la restriction de p_H au cône $\mathbf{C}$ des fonctions continues et gardons la notation p_H pour sa restriction au cône $\mathbf{C}$ des fonctions boréliennes. Une application standard du théorème de Hahn-Banach montre que l'ensemble

$$H_c = \{m\in E^{\ddagger} : \forall f \in \mathbf{C} \quad m(f) \leqq p_H^c(f)\}$$

est l'enveloppe convexe, héréditaire (i.e. $\lambda \in H_c$ et $\mu \leqq \lambda \Rightarrow \mu \in H_c$) et fermée pour la topologie vague de H. Nous citons ici notre topologie métrisable compacte habituelle sur $E^{\ddagger}$ car, bientôt, nous ferons intervenir la topologie bien plus fine de la norme (définie par exemple par la distance $d(\lambda,\mu) = \sup\{|\lambda(f) - \mu(f)|;\ f \in \mathbf{C},\ 0 \leqq f \leqq 1\}$). Maintenant, que peut-on dire de l'ensemble, défini de manière analogue

$$H_b = \{m\in E^{\ddagger} : \forall f \in \mathbf{B} \quad m(f) \leqq p_H(f)\}$$

bien plus petit que H_c en général? Avant d'y répondre, rappelons qu'une partie B de $E^{\ddagger}$ (ou, plus généralement, d'un convexe compact métrisable plongé dans un e.l.c.) est dite *fortement convexe* si l'enveloppe convexe fermée de tout compact contenu dans B est encore contenue dans B. On définit de manière évidente la notion d'enveloppe fortement convexe, et on montre que l'enveloppe fortement convexe d'une partie analytique A de $E^{\ddagger}$ est l'ensemble des barycentres des mesures de probabilité sur $E^{\ddagger}$ qui sont portées par A. Revenons à notre question et, suivant Mokobodzki, posons pour H partie variable de $E^{\ddagger}$ et $m \in E^{\ddagger}$ fixée

$$I(H) = \inf_{f\in\mathbf{B},\ 0\leqq f\leqq 1} \{p_H(f) + m(1-f)\}$$

On a $0 \leqq I(H) \leqq m(1)$ et on vérifie aisément qu'on a $m \in H_b$ ssi $I(H) = m(1)$. Le lemme suivant est alors le clé pour élucider la structure de H_b.

LEMMA. *Pour* m *fixée, la fonction d'ensemble* I *est une capacité et, pour* H *analytique, fortement convexe, on a*

$$I(H) = \sup_{\lambda \in H} (\lambda \wedge m)(1)$$

DEMONSTRATION. Nous ne donnerons que les grandes lignes de la démonstration, mais avec suffisamment de détails pour que le lecteur puisse la reconstituer. Pour abréger, nous noterons $\mathbf{B}^1$ (resp $\mathbf{C}^1$) la partie convexe de $\mathbf{B}$ (resp $\mathbf{C}$) constituée des f comprises entre 0 et 1. Voyons d'abord la montée de I. Soient $H_n \uparrow H$ et $t > \lim_n I(H_n)$, et soit pour chaque n $f_n \in \mathbf{B}^1$ telle que $p_{H_n}(f_n) + m(1-f_n) < t$. Par extractions de sous-suites, et en utilisant le fait qu'adhérence forte et faible coincident pour un convexe, on se ramène successivement au cas où (f_n) converge faiblement dans $L^1(m)$, puis fortement dans $L^1(m)$, puis finalement au cas où (f_n) converge m-p.p.. Posant $f = \liminf f_n$, on a alors

$$m(1-f) = \lim_n m(f_n), \quad p_{H_k}(f) \leqq \liminf_n p_{H_k}(f_n) \text{ \textit{pour tout} } k\,,$$

l'inégalité de droite provenant du fait que nos formes sous-linéaires sont montantes. On en déduit sans peine qu'on a $I(H) \leqq t$, d'où la montée. Pour la descente sur les compacts, on remarque d'abord que $(H,f) \to p_H(f)$ est une capacité à deux arguments (cf th. 7); ses propriétés de descente entrainent alors qu'on a, pour H compact,

$$I(H) = \inf_{f \in \mathbf{C}^1} \{p_H(f) + m(1-f)\}$$

(où $\mathbf{C}^1$ a remplacé $\mathbf{B}^1$), puis la descente de I sur les compacts. Enfin, pour démontrer l'égalité de l'énoncé, on commence par traiter le cas où H est un convexe compact. Le premier membre I(H) vaut alors

$$\inf_{f\in\mathbf{C}^1} \sup_{\lambda\in H} \{\lambda(f) + m(1-f)\}$$

tandis que le second vaut

$$\sup_{\lambda\in H} \inf_{f\in\mathbf{C}^1} \{\lambda(f) + m(1-f)\}$$

d'après un calcul classique de la masse de $\lambda \wedge m$. L'égalité des deux provient alors du théorème du minimax (un avatar du théorème de Hahn-Banach). Pour terminer, i.e. pour avoir l'égalité de l'énoncé quand H est fortement convexe, analytique, il n'y a plus qu'à appliquer le théorème de capacitabilité à H et I.

Voici alors le résultat final de Mokobodzki; sa portée nous semble devoir dépasser l'utilisation que nous en ferons en théorie du balayage.

THEOREME 13. *Si* H *est une partie analytique de* $E^{\ddagger}$, *alors l'ensemble*

$$H_b = \{m\in E^{\ddagger}: \forall f \in \mathbf{B} \quad m(f) \leqq p_H(f)\}$$

est l'enveloppe fortement convexe, héréditaire, et fermée pour la topologie de la norme, de H. *De plus, on peut remplacer "fermée pour la topologie de la norme" par "fermée pour les limites de suites croissantes".*

DEMONSTRATION. Le dernier point, un peu surprenant a priori, est laissé à la sagacité du lecteur; en fait, l'adhérence pour la norme de toute partie héréditaire de $E^{\ddagger}$ est l'ensemble des limites des suites croissantes contenues dans cette partie. Passons au point principal et notons H' l'enveloppe en question de H. D'abord, il est clair que H_b est fortement convexe, héréditaire, et fermé pour la norme: il contient donc H'. Réciproquement, fixons $m \in H_b$ et soit I la capacité associée à m comme ci-dessus. Nous avons déjà signalé qu'on a $I(H) = m(1)$ et donc a fortiori $I(H') = 1$. Maintenant, H étant analytique, un calcul un peu long mais sans difficultés majeures montre que H' est analytique si bien qu'on peut appliquer le lemme. On trouve ainsi dans H' une suite (λ_n) telle que la masse de $\lambda_n \wedge m$ tende vers celle de m et donc telle que la suite des $\lambda_n \wedge m$ tende vers m pour la norme. Mais, H étant héréditaire et fermé pour la norme, les $\lambda_n \wedge m$ lui appartiennent ainsi que la limite m. Par conséquent, H' contient H_b, et c'est terminé.

COROLLAIRE. *Si* H *est analytique,* H_b *l'est encore.*

Nous pouvons maintenant terminer notre étude du balayage. Encore une fois, nous ne ferons qu'esquisser les démonstrations en évitant en particulier d'entrer sérieusement dans la technique des ensembles analytiques.

Si S est un opérateur sous-linéaire analytique comme nos P, Q, R et si λ, μ sont deux sous-probabilités, nous noterons λS la forme sous-linéaire $f \to \lambda^*(Sf)$ et nous dirons qu'on a $\lambda S \geqq \mu$ si on a $\lambda(Sf) \geqq \mu(f)$ pour toute f analytique, ou borélienne, ou s.c.s - on sait que cela revient au même. La proposition suivante, intéressante en elle-même, va nous fournir l'analyticité du balayage.

THEOREME 14. *Soit* J *une maison de jeu analytique.*

1) *L'ensemble* $J' = \{(x,m) : \varepsilon_x P_J \geqq m\}$ *est une maison analytique et c'est la plus grand maison admettant même operatéur* P *que* J.

2) *Plus généralement, l'ensemble* $L' = \{(\lambda,\mu) : \lambda P_J \geqq \mu\}$ *est une partie analytique de* $E^{\ddagger} \times E^{\ddagger}$.

DEMONSTRATION. Nous commençons par 1) qui est plus facile. D'après le théorème précédent, J'_x est pour tout $x \in E$ l'enveloppe fortement convexe héréditaire, fermée pour la norme, de J_x, et est donc analytique. Maintenant, si on regarde (ce que nous ne ferons pas!) comment on établit cette analyticité, on voit de manière immédiate que le paramètre x ne complique rien, d'où l'analyticité globale de J'. Le reste du point 1) est évident. Pour établir 2) de la même manière, il nous manque pour L' ce qu'était J pour J', à savoir une partie analytique L de $E^{\ddagger} \times E^{\ddagger}$ telle que, pour λ fixée (dans le premier facteur), on ait $\lambda P_J = p_{L\lambda}$ (où $p_{L\lambda}$ est la forme sous-linéaire associée à L_λ). Qu'une telle partie analytique L existe résulte (laborieusement) du fait que $P(\cdot,\cdot)$ est capacitaire (cf th. 7) et du théorème de capacitabilité précisé (évoqué après l'énoncé du corollaire du théorème 9).

COROLLAIRE. 1) *La saturée* $\tilde{J}$ *d'une maison analytique* J *est analytique.*

2) *Plus généralement, la relation de balayage d'une maison analytique* J *est (à graphe) analytique.*

DEMONSTRATION. D'après le corollaire du théorème 9, R_J est de la forme $P_{\check{J}}$ avec $\check{J}$ analytique, et on applique alors le théorème à $\check{J}$.

Dans l'énoncé suivant, qui est notre théorème à la Strassen, l'expression "N majoré par P" signifie évidemment "$Nf \leqq Pf$ pour toute f analytique, ..." ou encore "$\varepsilon_x N \leqq \varepsilon_x P$ pour tout x".

THEOREME 15. *Soient* J *une maison analytique et* $\lambda, \mu \in E^{\ddagger}$. *Pour que l'on ait* $\lambda P_J \geqq \mu$ *il faut et il suffit qu'il existe une suite* (N_k)

de noyaux sousmarkoviens majorés par P_J *telle que les mesures* λN_k *tendent en croissant vers* μ.

DEMONSTRATION. La condition suffisante est triviale. La nécessité se démontre en suivant un chemin analogue à celui emprunté pour le 2) du th. 14 (quoiqu'il ne s'agisse pas du même type de propriété). Grâce au th. 7 et au théorème de capacitabilité précisé, on exhibe (avec du travail) une partie analytique H de $E^{\ddagger}$ vérifiant les propriétés

- toute $\nu \in H$ est de la forme λN avec N noyau majoré par P_J
- la forme p_H est égale à λP_J.

Et on peut remplacer H par son enveloppe fortement convexe héréditaire, qui a clairement les mêmes propriétés et qui, par ailleurs, est encore analytique. Le théorème 13 implique alors que l'ensemble $\{\nu : \lambda P_J \geqq \nu\}$ est l'adhérence pour la norme de cette enveloppe, d'où la conclusion.

REMARQUE. On peut montrer que la suite (N_k) de l'énoncé admet une espèce de valeur d'adhérence N vérifiant $\mu = \lambda N$ et $Nf \leqq P_J f$ λ-p.p. pour toute $f \in \mathbf{B}$ (le "λ-p.p." dépendant de f). Mais il se peut qu'il n'existe aucun noyau M vérifiant $\mu = \lambda M$ et $\forall x\ \varepsilon_x M \leqq \varepsilon_x P_J$, même dans le cadre du corollaire suivant.

COROLLAIRE. *Soient* J *une maison analytique et* $\lambda, \mu \in E^{\ddagger}$. *Pour que l'on ait* $\lambda \dashv_J \mu$ *il faut et il suffit qu'il existe une suite* (N_k) *de noyaux sous-markoviens telle que* $\mu = \lim \uparrow \lambda N_k$ *et que l'on ait* $\varepsilon_x \dashv_J \varepsilon_x N_k$ *pour tout* x *et tout* k.

DEMONSTRATION. Ici encore on applique le théorème à la maison analytique $\check{J}$ que nous fournit le corollaire du théorème 9.

REMARQUE. Reprenons, dans le cadre du corollaire, le noyau N de la remarque précédente. Il vérifie ici $\mu = \lambda N$ et $Nf \leq f$ λ-p.p. pour toute $f \in \mathcal{S}_b$. Si le balayage est séparable, on voit aisément qu'on peut faire disparaitre ce "λ-p.p.", et on retrouve alors le th. 11.

Bibliographie

Nous renvoyons une fois encore le lecteur au 3ème volume de "Probabilités et Potentiel" pour des commentaires et une bibliographie sérieuse. Je me contenterai de dire ici que, si la théorie des opérateurs capacitaires me préoccupe depuis une dizaine d'années (en témoigne bon nombre de volumes du Séminaire de Probabilités de Strasbourg), les premières applications explicites à la théorie du balayage apparaissent dans un exposé - extrèmement dense - au Séminaire Choquet (1980/81 - 20e année - paru en 1982) tandis que la version élaborée apparait pour la première fois dans le volume susdit.

Claude DELLACHERIE
Département de Mathématique
Université de Rouen
B.P. n°67
76130 MONT SAINT AIGNAN
FRANCE

Seminar on Stochastic Processes, 1983
Birkhäuser, Boston, 1984

LOCAL TIMES AND QUANTUM FIELDS*

by

E.B. DYNKIN

The central point of the paper is an isomorphism theorem which establishes a relation between a Gaussian random field associated with a symmetric Markov process (the free field) and local times for the process. The free field associated with the Brownian motion plays an important role in constructive quantum field theory. The isomorphism theorem allows one to express moments of the cutoff $P(\phi)_2$ fields in terms of multiple local times for the Brownian motion. On the other hand, techniques of field theory can be applied to investigate local times and self-crossing properties of Markov processes.

The idea that local times and self-crossings of the Brownian motion can be used as a tool in quantum field theory is due to Symanzik [8]. Brydges, Fröhlich and Spencer [1] have applied self-crossings of a discrete Markov chain to classical spin systems. In [3] the same systems have been studied using local times for Markov processes with continuous time parameter.

*Research supported in part by NSF Grant MCS-8202286.

1. Local times

If X_t is a stochastic process in a space E, then the integral

$$\int_I \delta_z(X_t)dt \tag{1.1}$$

describes the time spent by a particle at point $z \in E$ during the time interval I. Here δ_z is "the delta-function at the point z." If E is discrete and if X_t is a Markov process with symmetric transition probabilities $p_t(x,y) = p_t(y,x)$, then $\delta_z(x) = 1_z(x)$ is equal to 1 for $x = z$ and vanishes for $x \neq z$. In general (1.1) is defined as a certain limit. The transition density

$$p_s(z,x) = (2\pi s)^{-d/2} \exp\{-\frac{1}{2s}|x - z|^2\} \tag{1.2}$$

of the Brownian motion in $\mathbb{R}^d$ can be considered as an approximation of $\delta_z(x)$. In the case $d = 1$,

$$\lim_{s\to 0} \int_I p_s(z,X_t)dt \tag{1.3}$$

exists in L^2 for every finite interval $I \subset \mathbb{R}_+$ and it can be chosen to be an additive functional of X_t. The value of $\delta_z(X_t)$ is not defined but the integral (1.1) is defined by (1.3). We consider $\delta_z(X_t)$ as a generalized function of t. For $d \geq 2$, it is a generalized function of both t and z. Instead of (1.1) and (1.3) we consider

$$\int_E \lambda(dz) \int_I \delta_z(X_t)dt = \lim_{s\to 0} \int_E \lambda(dz) \int_I p_s(z,X_t)dt. \tag{1.4}$$

An L^2-limit exists for every measure λ such that

(1.5) $$\int_{E\times E} \lambda(dx)\,\lambda(dy)\, g^u(x,y) < \infty \quad \text{for all finite } u$$

where

(1.6) $$g^u(x,y) = \int_0^u p_t(x,y)dt.$$

If λ has a density with respect to the Lebesgue measure m, then

(1.7) $$\int_E \lambda(dz) \int_I \delta_z(X_t)dt = \int_I \frac{d\lambda}{dm}(X_t)\, dt.$$

The same is true for every fine Markov process with a symmetric transition density, and only slight modifications are needed in the case of a symmetric transition function which has no density (see [2]).

We assume that Green's function

(1.8) $$g(x,y) = \int_0^\infty p_t(x,y)dt$$

is finite for $m\times m$ —almost all x,y. This condition is satisfied for the Brownian motion in $\mathbb{R}^d$ with a constant killing rate r (which means that a factor e^{-rs} should be added in formula (1.2)) or for the Brownian motion killed on the boundary of a bounded domain.

The occupation field

(1.9) $$T_z = \int_{\mathbb{R}_+} \delta_z(X_t)dt, \qquad z \in E$$

is a generalized random field: a value

$$T_\lambda = \int_E \lambda(dz)T_z$$

is defined for every measure λ such that

$$(1.10) \qquad \int_{E\times E} \lambda(dx)\, \lambda(dy)\, g(x,y) < \infty.$$

(Since the process X_t terminates at a finite time ζ, the integral (1.9) should be interpreted as an integral from 0 to ζ; formally, we put $\delta_z(X_t) = 0$ for $t \geq \zeta$.)

2. Multiple local times

Time-space location of self-crossings of a path can be described by the integrals

$$(2.1) \qquad \int_{I_1} dt_1 \cdots \int_{I_n} dt_n\, \delta_z(X_{t_1}) \cdots \delta_z(X_{t_n}), \qquad n = 1,2,\ldots,\ z \in E,$$

$$I_1,\ldots, I_n \quad \text{are open intervals.}$$

For discrete case, the integrals (2.1) are well defined. In general, it is natural to consider an L^2-limit

$$(2.2) \qquad \int_E \lambda(dz) \int_{I_1} dt_1 \cdots \int_{I_n} dt_n\, \delta_z(X_{t_1}) \cdots \delta_z(X_{t_n})$$

$$= \lim_{s\to 0} \int_E \lambda(dz) \int_{I_1} dt_1 \cdots \int_{I_n} dt_n\, p_s(z,X_{t_1}) \cdots p_s(z,X_{t_n}).$$

Formula (2.2) for $n = 1$ is identical with (1.4). The limit exists if

$$(2.3) \qquad \int_{E\times E} \lambda(dx)\, \lambda(dy)\, g^u(x,y)^n < \infty$$

for all finite u and if the intervals $I_1,\ldots,I_n$ are disjoint and have no common ends. Condition (1.5) is a particular case of (2.3). All finite measures with bounded densities have property (2.3) if X_t is a Brownian motion in $\mathbb{R}^d$ and:

(a) $d \le 2$, n is arbitrary, or

(b) $n = 1$, d is arbitrary, or

(c) $d = 3$, $n = 2$.

To every λ, subject to condition (2.3), there corresponds a random measure L_λ on $\mathbb{R}_+^n$ such that $L_\lambda(I_1 \times \cdots \times I_n)$ coincides a.s. with the limit (2.2) for every disjoint interval $I_1, \ldots, I_n$ without common ends. This measure charges no hyperplane $t_i = \text{const}$ and no hyperplane $t_i = t_j$. If $d > 1$, $L_\lambda(B) = \infty$ for every open rectangle $I_1 \times \cdots \times I_n$ which intersects any of the planes $t_i = t_j$. However it is possible "to compensate infinities" and to define a random field

$$(2.4) \qquad :T^n:_\lambda = \int_E \lambda(dz) \int_{\mathbb{R}_+^n} :\delta_z(X_{t_1}) \cdots \delta_z(X_{t_n}): dt_1 \cdots dt_n$$

indexed by measures λ such that

$$(2.5) \qquad \int_{E\times E} \lambda(dx)\, \lambda(dy)\, g(x,y)^n < \infty.$$

A particular case

$$:T^2:_m = \int_E m(dz) \int_{\mathbb{R}^2} :\delta_z(X_s)\, \delta_z(X_t): ds dt = \int_{\mathbb{R}_+^2} :\delta(X_t - X_s): ds dt$$

has been studied by Varadhan [7].

3. Measures P_{xy}

We consider measures on the space W of paths defined by the following finite-dimensional distributions:

$$P_{xy}\{X_{t_1} \in B_1, \ldots, X_{t_n} \in B_n\} =$$

$$= \int_{B_1} \cdots \int_{B_n} p_{t_1}(x,z_1)\, m(dz_1)\, p_{t_2-t_1}(z_1,z_2)\, m(dz_2) \cdots$$

$$p_{t_n-t_{n-1}}(z_{n-1},z_n)\, m(dz_n)\, g(z_n,y)$$

for $t_1 < t_2 < \cdots < t_n$. Heuristically,

$$P_{xy}(\cdot) = g(x,y)\ \mathit{Probability}\ (\cdot\,|X_0 = x, X_{\zeta-} = y).$$

(Properly interpreted, this formula can be rigorously proved.) For any measures μ, ν on E we put

$$P_{\mu\nu} = \int_{E\times E} P_{xy}\, \mu(dx)\, \nu(dy).$$

Note that $P_{xy}(W) = g(x,y)$ and therefore

$$P_{\mu\nu}(W) = \int_{E\times E} \mu(dx)\, g(x,y)\, \nu(dy).$$

4. Free field

Let $g(x,y)$ be Green's function of a symmetric Markov process X_t. If $g(x,x)$ is finite for all x, then there exists a Gaussian random field ϕ_x, $x \in E$ such that

$$\langle\phi_x\rangle = 0, \quad \langle\phi_x\phi_y\rangle = g(x,y).$$

Here $\langle F\rangle$ means the mathematical expectation of F, i.e.,

$$\langle F\rangle = \int_\Omega F(\omega)\, \Pi(d\omega)$$

where (Ω,Π) is a probability space on which the field ϕ is defined (it has no relation to the path space W). We call ϕ *the free field associated with* X_t. It can be constructed even if $g(x,x) = \infty$. In this case ϕ_x is not defined but

$$\phi_\lambda = \int_E \phi_x \lambda(dx)$$

can be defined for all measures λ subject to the condition (1.10): ϕ_λ is a Gaussian family such that

$$<\phi_\lambda> = 0, \quad <\phi_\lambda \phi_\mu> = \int_{E\times E} \lambda(dx)\, g(x,y)\, \mu(dy).$$

5. Isomorphism theorem. Case $g(x,x) < \infty$.

Let

$$\xi_x = \phi_x^2/2. \tag{5.1}$$

(This formula cannot be used if $g(x,x)$ is infinite. A definition of the field ξ_x which is applicable to the general case will be given in Section 7.)

THEOREM 1. *The fields* $(\xi.(\omega), \phi_\mu^2(\omega)\, \Pi(d\omega))$ *and* $(\xi.(\omega) + T.(w), P_{\mu\mu}(dw)\, \Pi(d\omega))$ *are identical in distribution. In other words, for every positive functional* F *of* ξ,

$$<\phi_\mu^2 F(\xi)> = P_{\mu\mu}<F(\xi + T)>. \tag{5.2}$$

It is sufficient to check (5.2) for $F(\xi) = e^{-\xi_\lambda}$ where

$$\xi_\lambda = \int_E \xi_x \lambda(dx)$$

and λ satisfies condition (1.10). Indeed, if (5.2) holds for the functionals of this form, then it holds for every positive F measurable with respect to the σ-algebra generated by ξ_x. We consider a Markov process $\hat{X}_t$ obtained from X_t by killing with the killing measure λ. The free field associated with $\hat{X}_t$ is identical in distribution with $(\phi, \hat{\Pi})$ where

$$\hat{\Pi}(C) = \langle 1_C \, e^{-\xi_\lambda} \rangle / \langle e^{-\xi_\lambda} \rangle$$

and Green's function for $\hat{X}_t$ equals $P_{xy} e^{-T_\lambda}$.

Formula (5.2) follows also from the relation

$$(5.3) \quad \langle F(\xi) \exp(\phi_\mu - \tfrac{1}{2} P_{\mu\mu}(W)) \rangle = \int F(\xi(\omega) + T(w_1) + \cdots + T(w_N)) dM_\mu .$$

Here M_μ is a probability measure on the space of configurations in W such that

$$\int f(w_1, \ldots, w_N) dM_\mu = \sum_{n=0}^{\infty} \frac{1}{n!} e^{-Q(W)} \int_{W^n} f(w_1, \ldots, w_n) \, Q(dw_1) \cdots Q(dw_n)$$

where $Q = P_{\mu\mu}/2$ (that is M_μ corresponds to the Poisson point process in W with the characteristic measure $P_{\mu\mu}/2$).

For the case of a finite space E, (5.2) and (5.3) have been proved in [3].

6. Application to local times

It follows from Theorem 1 that, if H is a linear space of functions from E to $\mathbb{R}$ and if $\xi_\cdot(\omega)$ belongs to H for Π-almost all ω, then $T(w)$ belongs to H for P_{xy}-almost all w (for every x,y). Indeed, if $\xi(\omega) \in H$ Π-a.s., then $\xi(\omega) \in H$ $\phi_x \phi_y \Pi$-a.s. and, by the isomorphism theorem, $\xi(\omega) + T(w) \in H$ $\Pi \times P_{xy}$-a.s. By Fubini's theorem, for Π-almost all ω, $\xi(\omega)+T(w) \in H$ P_{xy}-a.s. Hence there exists ω such that $\xi(\omega) \in H$ and $\xi(\omega)+T(w) \in H$ for P_{xy}-almost all w, which implies that $T(w) \in H$ P_{xy}-a.s.

As an example, we consider the Brownian motion in $\mathbb{R}^1$ with constant killing rate 1/2. Green's function $g(x,y) = e^{-|y-x|}$, and the corresponding free field is the Ornstein-Uhlenbeck process. Its paths are continuous and so are paths of $\xi_x = \phi_x^2/2$. Hence paths of T_x are continuous P_{xy}-a.s.

By applying formula (5.2) to $F(\xi) = \exp[-\sum \lambda_i \xi_{x_i}]$ we get the following expression

$$P_{xy} \exp(-\sum \lambda_i T_{x_i}) = \frac{\langle \phi_x \phi_y \exp(-\sum \lambda_i \xi_{x_i}) \rangle}{\langle \exp(-\sum \lambda_i \xi_{x_i}) \rangle} .$$

P. Sheppard has shown that results of Knight [4,5] and Ray [7] on local times for one-dimensional diffusions can be deduced from this expression.

7. Isomorphism theorem. General case.

Now we replace the condition $g(x,x) < \infty$ by a less restrictive assumption

$$\int_s^\infty p_t(x,x)dt < \infty \quad \text{for} \quad s > 0 \tag{7.1}$$

(which is satisfied, in particular, for the killed Brownian motion in $\mathbb{R}^d$ for all d).

For every $s > 0$ and every x, the measure $\lambda(dy) = p_s(x,dy)$ satisfies condition (1.10) and therefore

$$\phi_{sx} = \int_E p_s(x,y)\, m(dy)\, \phi_y \tag{7.2}$$

is well-defined for all $s > 0$ and all x. If

$$\int_{E\times E} \lambda(dy)\, g(x,y)^2\, \lambda(dy) < \infty, \tag{7.3}$$

then there exists

$$\xi_\lambda = \lim_{s\to 0} \int_E \lambda(dx)(\phi_{sx}^2/2 - \langle\phi_{sx}^2/2\rangle). \tag{7.4}$$

Let Λ be the class of all measures with properties (1.10) and (7.3). Theorem 1 holds for all positive functionals of the field ξ_λ, $\lambda \in \Lambda$, i.e., for all positive functions on Ω which are measurable with respect to the σ-algebra generated by ξ_λ, $\lambda \in \Lambda$. (Every such function has the form $F(\xi_{\lambda_1},\ldots,\xi_{\lambda_n},\ldots)$ for some $\lambda_1,\ldots,\lambda_n,\ldots \in \Lambda$).

In the particular case considered in Section 5

$$\xi_\lambda = \int_E \xi_x \lambda(dx), \qquad \xi_x = \phi_x^2/2 - \langle\phi_x^2/2\rangle.$$

(Actually Theorem 1 holds in this case for every $\xi_x = \phi_x^2/2 - a(x)$ with an arbitrary measurable function $a(x)$.)

8. Wick's powers of the free field and powers of the field ξ

Wick's powers of a Gaussian random variable Y with mean 0 are defined by the following generating function

$$\sum_{n=0}^\infty \frac{t^n}{n!} :Y^n: = \exp(tY - \tfrac{1}{2} t^2 \langle Y^2\rangle).$$

In particular $:Y^0: = 1$, $:Y^1: = Y$, $:Y^2: = Y^2 - \langle Y^2\rangle$.

If

$$\int_{E\times E} \lambda(dx)\, g(x,y)^n\, \lambda(dy) < \infty, \tag{8.1}$$

then there exists an L^2-limit

$$U_\lambda^n = \lim_{s\to 0} \int_E :\phi_{sx}^n: \lambda(dx). \tag{8.2}$$

Note that

$$U_\lambda^0 = \lambda(E),\ U_\lambda^1 = \phi_\lambda,\ U_\lambda^2 = 2\xi_\lambda.$$

It is customary to write

$$U_\lambda^n = \int_E :\phi_x^n: \lambda(dx) \tag{8.3}$$

and to call the generalized field $:\phi_x^n:$ *the* n*th Wick's power of the free field.*

In the rest of the paper we assume that X_t is an exponentially killed Brownian motion in $\mathbb{R}^2$, i.e. that

$$p_s(x,y) = (2\pi s)^{-1} \exp(-ks - \frac{1}{2s}|x-y|^2), \quad k > 0. \tag{8.4}$$

Condition (8.1) is satisfied for all n if λ is a finite measure with a bounded density. We call such measures admissible.

THEOREM 2. *For every admissible* λ *and every* $n = 0,1,2,\ldots,$

$$V_\lambda^n = 2^{-n}\, U_\lambda^{2n}$$

is a functional of the field ξ. *Namely, let*

$$\xi_{sx} = \int_E p_s(x,dy)\xi_y. \tag{8.5}$$

Then

$$V_\lambda^n = \lim_{s\to 0} \int_E \lambda(dx) \sum_{k=0}^n b_{nk}^s(x)\xi_{sx}^k \tag{8.6}$$

(convergence in quadratic mean). Here $b^s_{nn} = 1$ *and* b^s_{nk} *are polynomials of chain variables*

$$(8.7) \qquad C^s_n(x) = \int_{E^n} p_s(x,dy_1)\cdots p_s(x,dy_n)\; g(y_1,y_2)\cdots g(y_{n-1},y_n)$$

and loop variables

$$(8.8) \qquad L^s_n(x) = \frac{1}{2n}\int_{E^n} p_s(x,dy_1)\cdots$$

$$p_s(x,dy_n)\; g(y_1,y_2)\cdots g(y_{n-1},y_n)\; g(y_n,y_1).$$

More precisely, the countable triangular matrix (b_{nk}) *is the inverse for a matrix* (a_{nk}) *which is defined by the following formula*

$$(8.9) \qquad \sum_{0\le k\le n<\infty} \frac{a_{nk}}{n!} u^n v^k = \exp(v\mathfrak{C}(u) + \mathfrak{L}(u)).$$

Here

$$(8.10) \qquad \mathfrak{C}(u) = u + \sum_{n=2}^{\infty} C_n u^n, \quad \mathfrak{L}(u) = \sum_{n=2}^{\infty} L_n u^n$$

are the generating functions of the chain variables and the loop variables respectively.

We write

$$(8.11) \qquad V^n_\lambda = \int_E :\xi^n_x: \lambda(dx)$$

and we call the generalized field ξ^n_x the nth power of ξ.

9. $P(\phi)_2$ fields

Suppose that Y is a functional of the free field such that

$$(9.1) \qquad \langle e^{-Y}\rangle < \infty.$$

The formula

$$\Pi_Y(F) = Z^{-1} \langle Fe^{-Y} \rangle \tag{9.2}$$

(Z is a normalizing constant) defines a probability measure on Ω which we call the Y-perturbation of the measure Π. Let λ be an admissible measure and let $P(s) = a_0 s^n + a_1 s^{n-1} + \cdots + a_n$ be a polynomial bounded from below. Nelson [6] has proved that condition (9.1) holds for

$$Y_\lambda = a_0 :\phi^n:_\lambda + a_1 :\phi^{n-1}:_\lambda + \cdots + a_n .$$

We call the corresponding perturbation Π_{Y_λ} a *cutoff* $P(\phi)_2$ *measure*. (Subscript refers to the dimension of the index space $E = \mathbb{R}^2$.) Let m^r be the restriction of the Lebesgue measure m to the ball $|x| \le r$. The $P(\phi)_2$ *measure* (without cutoff) can be constructed as the limit of $\Pi_{Y_{m^r}}$ as $r \to \infty$.

Fields corresponding to monomials $P(s) = as^{2n}$ are called $(\phi^{2n})_2$ *fields*. The expectation for the cutoff $(\phi^{2n})_2$ field is given by the formula

$$\langle\langle F \rangle\rangle = Z^{-1} \langle e^{-cV_\lambda^n} F \rangle, \quad c = 2^n a.$$

By Theorem 1

$$\langle\langle \phi_x \phi_y \rangle\rangle = Z^{-1} P_{xy} \langle e^{-cV_\lambda^n(\xi + T)} \rangle .$$

Similar expressions can be written for higher moments and for cutoff $P(\phi)_2$ fields.

10. Powers of the occupation field

THEOREM 3. *Let*

$$T_{sx} = \int_E p_s(x,dy) T_y = \int_0^\zeta p_s(x,X_t)dt. \tag{10.1}$$

For every n *and every admissible measure* λ, *there exists a random variable* $:T^n:_\lambda$ *such that*

$$(10.2) \qquad :\frac{T^n}{n!}:_\lambda = \lim_{s\to 0} \int_E \lambda(dx) \sum_{k=0}^{n} B^s_{nk}(x)\, T^k_{sx}/k!$$

in $L^2(P_{\mu\mu})$ *for all admissible measures* μ. *Here* B_{nk} *are polynomials of the chain variables with the generating function*

$$(10.3) \qquad \sum_{n=k}^{\infty} B_{nk}\, u^n = \mathfrak{D}(u)^k, \quad k = 0,1,\ldots$$

where $\mathfrak{D}(u)$ *is the inverse function for* $\mathfrak{C}(u)$.

To prove Theorem 3, we note that by Theorems 1 and 2 there exists an $L^2(\Pi \times P_{\mu\mu})$-limit of

$$\int_E \lambda(dx) \sum_{k=0}^{n} b^s_{nk}(x)(\xi_{sx} + T_{sx})^k.$$

The integral of this expression with respect to Π converges in $L^2(P_{\mu\mu})$. Formula (10.3) follows from (8.9).

References

1. D. BRYDGES, J. FRÖHLICH and T. SPENCER. The random walk representation of classical spin systems and correlation inequalities. *Comm. Math. Phys. 83* (1982), 123-150.

2. E.B. DYNKIN. Green's and Dirichlet spaces associated with fine Markov processes. *J. Funct. Analysis, 47* (1982), 381-418.

3. E.B. DYNKIN. Markov processes as a tool in field theory. *J. Funct. Analysis, 50* (1983), 167-187.

4. F.B. KNIGHT. Random walks and a sojourn density process of

Brownian motion. *Trans. Amer. Math. Soc. 109* (1963), 56-86.

5. F.B. KNIGHT. Brownian local times and taboo processes. *Trans. Amer. Math. Soc. 143* (1969), 173-185.

6. E. NELSON. Probability theory and Euclidean field theory. In: *Constructive Quantum Field Theory*. G. Velo and A. Wightman, eds. Springer-Verlag, New York, 1973.

7. D. RAY. Sojourn times of diffusion processes. *Illinois J. Math. 7* (1963), 615-630.

8. K. SYMANZIK. Euclidean quantum field theory. Appendix by S.R.S. Varadhan. In: *Local Quantum Theory*, R. Jost, ed. Academic Press, New York/London, 1969.

E.B. DYNKIN
Department of Mathematics
Cornell University
Ithaca, New York 14853

ADDED IN PROOFS. Results presented in this paper are proved in detail in the following publications:

E.B. DYNKIN. Gaussian and non-Gaussian random fields associated with Markov processes. *J. Funct. Anal. 55* (1984), 344-376.

E.B. DYNKIN. Polynomials of the occupation field and related random fields. *J. Funct. Anal. 57* (1984).

Seminar on Stochastic Processes, 1983
Birkhäuser, Boston, 1984

APPROXIMATION OF DEBUTS*

by

NEIL FALKNER

ABSTRACT. Consider a Markov process X whose state space is an arbitrary measurable space not endowed with any topology to begin with. Assume only that X has no branch points and that for each α-excessive function f, the process $(f(X_t))$ is a.s. right continuous. Let $\mathcal{E}^e$ be the σ-field generated by the α-excessive functions. We show that for each $A \in \mathcal{E}^e$ and each finite measure μ on the state space, there is a decreasing sequence (G_n) of finely open, $\mathcal{E}^e$-measurable supersets of A such that $D_{G_n} \uparrow D_A$ P^μ-a.s. We deduce this result from a similar one which applies also to suitable non-Markov processes.

1. Introduction

Consider a Markov process X. For any set A, the *debut* of A is defined by $D_A = \inf\{t \geq 0: X_t \in A\}$. Under suitable hypotheses on X, the following statement holds:

(1.1) *For each nearly Borel set* A *and each finite measure* μ *on the state space, there is a decreasing sequence* (G_n) *of finely open nearly Borel sets* $G_n \supseteq A$ *such that* $D_{G_n} \uparrow D_A$ P^μ*-a.s.*

*Research supported in part by NSF grant MCS-8103473.

This result is technical in appearance but is well known to be of central importance in potential theory. It was first shown to hold for Hunt processes [7], then for standard processes [1, 2], and finally for right processes [12, 13, 6]. Perhaps the simplest proof of (1.1) for right processes is to be found in [10], but there, as in [12] but not in [13] or [6], the resolvant of X is assumed to map Borel functions to Borel functions, a property which is not preserved under certain procedures for forming new processes from old ones. However, as we shall show, the simple argument given in [10] may be adapted to prove the following result:

THEOREM 1.1. Let $(E,\mathcal{E})$ be an arbitrary measurable space and let $\mathcal{E}^*$ be the universal completion of $\mathcal{E}$. Let $X = (\Omega,M,(M_t)_{0 \leq t < \infty}, (X_t)_{0 \leq t < \infty}, (P^x)_{x \in E})$ be a Markov process with state space $(E,\mathcal{E}^*)$ and with cone of α-excessive functions S^α $(0 \leq \alpha < \infty)$. Let $S^\infty = \bigcup_\alpha S^\alpha$, let $\mathcal{E}^e = \sigma(S^\infty)$, and let $\mathcal{O}_f$ be the weakest topology on E such that each element of S^∞ is $\mathcal{O}_f$-continuous. Assume that:

(a) For each finite measure ν on $\mathcal{E}^*$, and each $f \in S^\infty$, the process $(f(X_t))$ is P^ν-a.s. right-continuous.

(b) For each $x \in E$, $\{x\} \in \mathcal{E}^*$, and $P^x(X_0 = x) = 1$.

Then for each $A \in \mathcal{E}^e$ and each finite measure μ on $\mathcal{E}^*$, there exists a decreasing sequence (G_n) in $\mathcal{O}_f \cap \mathcal{E}^e$ such that $G_n \supseteq A$ for all n, and $D_{G_n} \uparrow D_A$ P^μ-a.s. as $n \to \infty$.

We emphasize that in the above theorem, it is not assumed that X satisfies any special properties other than (a) and (b). In particular, we make no topological assumptions whatsoever on E, whereas even in [6] it is assumed that E is a topological space which is homeomorphic to a universally measurable subspace of a separable completely metrizable space, that $\mathcal{E}$ = Borel E, and that (X_t) is right-continuous with respect to the topology of E. We mention that in the setting of [6], $\mathcal{O}_f$ is the fine topology, $\mathcal{E}^e$ contains the Borel sets and is contained in the σ-field of nearly Ray-Borel sets, and if the α-excessive functions are nearly Borel, then the nearly Ray-Borel sets and the nearly Borel sets are the same.

Now (1.1) is a kind of measure-theoretic approximation result, and it seems odd that its proof should involve the Markov property of X. Indeed, the argument used to prove (1.1) in [7] depended on right-continuity and quasi-left-continuity of X but not on the Markov property. In Section 2 we prove a result like (1.1) for a process X which is not assumed to be Markov or quasi-left-continuous. When specialized to the setting of [7], our approach yields G_n's which are not just finely open but actually open as in [7]. In [12], [10], [13], and [6], the arguments used to prove (1.1) do not assume quasi-left-continuity of X but do make use of the Markov property and of the resolvent of X. In Section 3 we consider a Markov process X and use the result of Section 2 to prove a result even more general than Theorem 1.1.

2. General Processes

This section is rather abstract and the reader may find it helpful to skim over Section 3 first for motivation.

Let E be a set and let $\mathcal{U}$ be a collection of *bounded* real-valued functions on E. Let $\mathcal{E}^e = \sigma(\mathcal{U})$. Let $(\Omega,\mathcal{B},P)$ be a *complete* probability space, let $(\mathcal{B}_t)_{0 \leq t < \infty}$ be a filtration of $(\Omega,\mathcal{B},P)$ satisfying the *usual hypotheses* (see [4], p. 183), and let $(X_t)_{0 \leq t < \infty}$ be a stochastic process over $(\Omega,\mathcal{B},P)$ with values in $(E,\mathcal{E}^e)$ and *adapted* to $(\mathcal{B}_t)$. Assume that:

(2.1) *For each* $f \in \mathcal{U}$, *the process* $(f(X_t))$ *is a.s. right-continuous.*

From (2.1) it follows that for each $g \in b\mathcal{E}^e$, there exists $\Lambda \in \mathcal{B}$ such that $P(\Lambda) = 1$ and the process $(1_\Lambda\, g(X_t))$ is well measurable (with respect to $(\mathcal{B}_t)$). Hence if $A \in \mathcal{E}^e$, then D_A is a stopping time. A real-valued process (Z_t) will be called *conditionally quasi-left-continuous* (abbreviated CQLC) iff whenever (T_n) is an increasing sequence of stopping times with limit T, we have a.s.

$$\lim_n Z_{T_n} 1_{\{T < \infty\}} = E[Z_T 1_{\{T < \infty\}} \mid \bigvee_n \mathcal{B}_{T_n}] .$$

(Another term for CQLC is "regular" [11].) Assume that:

(2.2) *For each* $f \in \mathcal{U}$, *the process* $(f(X_t))$ *is* CQLC.

The class of CQLC processes does not have good stability properties. For example, in general the infimum of two CQLC processes is not CQLC. Let $\mathcal{L}$ be the smallest real vector sublattice of $b\mathcal{E}^e$ which contains $\mathcal{U} \cup \{1\}$. Assume that:

(2.3) *For each* $g \in \mathcal{L}$, *there is a uniformly bounded sequence* (g_m) *in* $\mathcal{U}$ *such that* $g_m \to g$ *pointwise on* E.

Now recall that a *semitopology* on E is collection $\mathcal{T}$ of subsets of E to which $\emptyset$ and E belong, and which is closed under finite intersections and countable unions. A function f from E into a second countable topological space M will be called $\mathcal{T}$-continuous iff $f^{-1}[V] \in \mathcal{T}$ for every open set $V \subseteq M$. Let $\mathcal{O}$ denote the smallest semitopology on E such that every $f \in \mathcal{U}$ is $\mathcal{O}$-continuous. Note that $\mathcal{O} \subseteq \mathcal{E}^e$.

THEOREM 2.1. Let $A \in \mathcal{E}^e$. Then there is a decreasing sequence (G_n) in $\mathcal{O}$ such that $G_n \supseteq A$ and $D_{G_n} \uparrow D_A$ a.s.

PROOF. *STEP 1.* Let M be the separable, completely metrizable space $\mathbb{R}^{\mathbb{N}\times\mathbb{N}}$. We are going to construct a certain map J from E into M. The image of E in M will not necessarily be measurable in any sense, but we shall show that it is contained in a Borel set F such that all the properties we need which hold on the image of E also hold on F.

Let H be a countable subset of U such that $A \in \sigma(H)$. We claim there is a countable rational vector sublattice $G \subseteq L$ such that $H \cup \{1\} \subseteq G$ and:

(2.4) *For all* $g \in G$, *there exists a uniformly bounded sequence* (g_m) *in* $G \cap U$ *such that* $g_m \to g$ *pointwise on* E.

Let G_0 be the smallest rational vector sublattice of L which contains $H \cup \{1\}$. Then G_0 is countable so by (2.3) there is a countable set $U_0 \subseteq U$ such that each element of G_0 is the pointwise limit of a uniformly bounded sequence in U_0. Let G_1 be the smallest rational vector sublattice of L which contains $G_0 \cup U_0$. Continue in this way, defining G_k and U_k for all $k \in \mathbb{N}$. Let $G = \bigcup_k G_k$. Then G has the desired properties. This establishes the claim.

Now let $(g_n)_{n\in\mathbb{N}}$ be a sequence whose range is G and for each n, choose a uniformly bounded sequence $(g_{nm})_{m\in\mathbb{N}}$ in $G \cap U$ such that $g_{nm} \to g_n$ pointwise on E as $m \to \infty$. Let $A = \sigma(g_{nm}\colon n,m \in \mathbb{N})$. Then $A = \sigma(G)$ and $A \in A$. Define $J\colon E \to M$ by $J(x) = (g_{nm}(x))_{(n,m)\in\mathbb{N}\times\mathbb{N}}$. Then:

(2.5) J *is 0-continuous.*

For $n,m \in \mathbb{N}$ let $\pi_{nm}\colon M \to \mathbb{R}$ be the (n,m)-th projection map, so $g_{nm} = \pi_{nm} \circ J$. Next, let $M' = \mathbb{R}^{\mathbb{N}}$ and define $J'\colon E \to M'$ by $J'(x) = (g_n(x))_{n\in\mathbb{N}}$ so for every n, $g_n = \pi_n \circ J'$ where $\pi_n\colon M' \to \mathbb{R}$ is the n-th projection map. Since $A = \sigma(G)$, we have $J(x_1) = J(x_2)$ iff $J'(x_1) = J'(x_2)$ so there is a (unique) map $\psi_o\colon J[E] \to J'[E]$ such that $\psi_o \circ J = J'$; moreover, ψ_o is a Borel isomorphism between $J[E]$ and $J'[E]$. Hence (see e.g. [4], p. 74) there are Borel sets $L \subseteq M$ and $L' \subseteq M'$ and a Borel isomorphism ψ of L onto L' such that $J[E] \subseteq L$, $J'[E] \subseteq L'$, and ψ extends ψ_o. Let $\rho_n = \pi_n \circ \psi$. Then $\rho_n \circ J = g_n$ and

$$(2.6) \qquad \sigma(\rho_n\colon\ n \in \mathbb{N}) = \text{Borel } L\,.$$

Now observe that there is a set $F \subseteq L$ such that:

$$(2.7) \qquad J[E] \subseteq F \in \text{Borel } M$$

and such that if we let $\gamma_n = \rho_n|F$ and $\Gamma = \{\gamma_n\colon\ n \in \mathbb{N}\}$, then:

(2.8) Γ *is a rational vector lattice of bounded functions on* F *with* $1 \in \Gamma$,

and

(2.9) $\pi_{nm} \to \gamma_n$ *pointwise and boundedly on* F *as* $m \to \infty$.

We have only to let F be the set of all $y \in L$ for which the following eight (not entirely independent) conditions hold for all $k,\ell,n \in \mathbb{N}$ and all rational numbers r:

(2.10) (a) $g_k = 0$ on $E \Rightarrow \rho_k(y) = 0$,

(2.10) (b) $g_k = 1$ on $E \Rightarrow \rho_k(y) = 1$,

(2.10) (c) $rg_k = g_\ell$ on $E \Rightarrow r\rho_k(y) = \rho_\ell(y)$,

(2.10) (d) $g_k + g_\ell = g_n$ on $E \Rightarrow \rho_k(y) + \rho_\ell(y) = \rho_n(y)$,

(2.10) (e) $|g_k| = g_\ell$ on $E \Rightarrow |\rho_k(y)| = \rho_\ell(y)$,

(2.10) (f) $|\rho_n(y)| \leq \sup_E |g_n|$;

(2.11) (a) $|\pi_{nk}(y)| \leq \sup_E |g_{nk}|$,

(2.11) (b) $\pi_{nm}(y) \to \rho_n(y)$ as $m \to \infty$.

Then (2.7) is clear, (2.8) follows from (2.10)(a) through (2.10)(f), and (2.9) follows from (2.11)(a) and (2.11)(b).

Now by (2.6),

$$\sigma(\Gamma) = \text{Borel } F . \tag{2.12}$$

Then by (2.8), (2.12), and a suitable version of the monotone class theorem,

(2.13) *The set of bounded Borel functions on* F *is equal to the smallest collection of functions on* F *which contains* Γ *and is closed under uniformly bounded pointwise limits of sequences.*

STEP 2. With the aid of the construction just described, we can complete the proof of the theorem along the lines of the proof of Théorème 2 of [10]. To convince the reader of this, we give the details. For $H \subseteq M$, let $R_H = D_{J^{-1}[H]}$. By (2.7) and the fact that $A \in \mathcal{A}$, there exists $B \in \text{Borel } M$ such that $J^{-1}[B] = A$ and $B \subseteq F$.

(a) First suppose B compact. Choose a decreasing sequence (V_p) of open subsets of M containing B such that $B = \bigcap_p \overline{V_p}$. Let $S_p = R_{V_p}$, $S = \uparrow \lim_p S_p$, and $S' = R_B$. We claim that $S = S'$ a.s. Evidently $S \leq S'$ and we have only to show that $J(X_S) \in B$ a.s. on $\{S < \infty\}$. By (2.3) and the definition of J, the limit $Y = \lim_p J(X_{S_p})$ exists in M a.s. on $\{S < \infty\}$, and for any bounded Borel function f on M, f(Y) is measurable with respect to $\bigvee_p \mathcal{B}_{S_p}$ so:

(2.14) $$E[\pi_{nm}(Y)f(Y);\ S < \infty] = E[\pi_{nm}(J(X_S))f(Y);\ S < \infty]\ .$$

Now by (2.1) and the definition of J, the process $(J(X_t))$ is a.s. right-continuous so on $\{S < \infty\} \subseteq \{S_p < \infty\}$ we have $J(X_{S_p}) \in \overline{V_p}$ a.s. Hence $Y \in B$ a.s. on $\{S < \infty\}$. In view of (2.9), since $B \subseteq F$, we find, by letting $m \to \infty$, that (2.14) holds with π_{nm} replaced by γ_n. Then, by (2.13), (2.14) holds with π_{nm} replaced by any bounded Borel function on F. But then, by a monotone class argument, on $\{S < \infty\}$ we have $Y = J(X_S)$ a.s., so indeed $J(X_S) \in B$ a.s. and the claim is established.

(b) Now consider the general case. For $H \subseteq M$ let $I(H) = E^*[e^{-R_H}]$ where E^* denotes "upper integral". It is automatic that $H_p \uparrow H$ implies $I(H_p) \uparrow I(H)$, and also that if $H_1, H_2 \subseteq M$ are open, then $I(H_1 \cup H_2) + I(H_1 \cap H_2) \leq I(H_1) + I(H_2)$. It follows (see [4], p. 89, (32.4)) that $I^*\colon\ \mathfrak{P}(M) \to [0,1]$ defined by

$$I^*(H) = \inf\{I(V):\ H \subseteq V \text{ open} \subseteq M\}$$

satisfies $I^*(H_p) \uparrow I^*(H)$ whenever $H_p \uparrow H$. Clearly, $I^*(H) = \inf\{I^*(V):\ H \subseteq V \text{ open} \subseteq M\}$ for all $H \subseteq M$. Therefore there are compact sets $C_k \subseteq B$ such that $I^*(C_k) \uparrow I^*(B)$. (See e.g. [3], Theorem 6.7.) But by (a), $I^*(C_k) = I(C_k)$. Therefore $I(B) = I^*(B)$. Thus there exists a decreasing sequence (V_p) of open subsets of M such that $B \subseteq V_p$ and $I(V_p) \downarrow I(B)$. Let $G_p = J^{-1}[V_p]$. Then $A \subseteq G_p$, $G_p \in \mathcal{O}$ by (2.5), and $D_{G_p} \uparrow D_A$ a.s. □

REMARK. In Theorem 2.1, the hypothesis that $A \in \mathcal{E}^e$ may be replaced by the weaker hypothesis that $A \in$ Souslin $\mathcal{E}^e$. Only minor changes in the proof are needed to establish this stronger version of the theorem.

3. Markov Processes

We first state our assumptions and prove the main result of this section. Then we explain how this result implies Theorem 1.1.

Let $(E, \mathcal{E}^e)$ be a measurable space and let $(U^\alpha)_{0 < \alpha < \infty}$ be a Markov resolvent ([8], p. 234) on $(E, \mathcal{E}^e)$. (For our purposes, the cemetery point, if there is one, may be treated like any other point of E, so there is no loss of generality in not considering sub-Markov resolvents.) Let $\mathcal{U} = \{U^\alpha f:\ f \in b\mathcal{E}^e\}$ (which is independent of $\alpha > 0$ by the resolvent equation). Assume that:

(3.0) $$\mathcal{E}^e = \sigma(\mathcal{U}).$$

Let $(\Omega, \mathcal{B}, (\mathcal{B}_t^o)_{0 \le t < \infty}, (X_t)_{0 \le t < \infty}, P)$ be a Markov process with state space $(E, \mathcal{E}^e)$, where $\mathcal{B}$ is complete with respect to P and $\mathcal{B}_o^o$ contains all P-null subsets of Ω, but $(\mathcal{B}_t^o)$ is not assumed to be right-continuous. Assume that:

(3.1) *For each* $f \in \mathcal{U}$, *the process* $(f(X_t))$ *is a.s. right-continuous.*

Then for each $g \in b\mathcal{E}^e$, the process $(g(X_t))$ is indistinguishable from a $(\mathcal{B}_t^o)$-progressively measurable process. (See e.g. [2], p. 34.) Assume that:

(3.2) *For each* $g \in b\mathcal{E}^e$, *each* $\alpha \in (0,\infty)$, *and each* $t \in [0,\infty)$,

$$E[\int_o^\infty e^{-\alpha s} g(X_s)ds \,|\, \mathcal{B}_t^o] = e^{-\alpha t} U^\alpha g(X_t) + \int_o^t e^{-\alpha s} g(X_s)ds .$$

The collection of functions $f \in b\mathcal{E}^e$ for which $\alpha U^\alpha f \to f$ pointwise on E as $\alpha \to \infty$ contains $\mathcal{U}$ but in general is not closed under lattice operations. Let $\mathcal{L}$ be the smallest real vector sublattice of $b\mathcal{E}^e$ which contains $\mathcal{U}$. (Note that $1 \in \mathcal{U}$.) Assume that:

(3.3) *For each* $g \in L$, $\alpha U^{\alpha} g \to g$ *pointwise on* E *as* $\alpha \to \infty$.

Let $\mathcal{O}$ be the smallest semitopology on E such that each $f \in \mathcal{U}$ is $\mathcal{O}$-continuous.

THEOREM 3.1. Let $A \in \mathcal{E}^e$. Then there is a decreasing sequence (G_n) in $\mathcal{O}$ such that $G_n \supseteq A$ and $D_{G_n} \uparrow D_A$ a.s.

PROOF. Clearly this is just the specialization of Theorem 2.1 to the setting of the present section. Now (2.1) and (2.3) hold by (3.1) and (3.3), respectively. It remains only to check that (2.2) holds. Let $f \in \mathcal{U}$ and let $g \in b\mathcal{E}^e$ such that $f = U^{\alpha} g$ for some $\alpha > 0$. By (3.2), the bounded process

$$Z_t = e^{-\alpha t} f(X_t) + \int_0^t e^{-\alpha s} g(X_s) ds$$

is a $(\mathcal{B}_t^o)$-martingale and by (3.1), it is a.s. right-continuous. Therefore it is a $(\mathcal{B}_t)$-martingale, where $(\mathcal{B}_t) = (\mathcal{B}_{t+}^o)$, and it is CQLC (with respect to $(\mathcal{B}_t)$). From this it follows easily that $(f(X_t))$ is CQLC; that is, (2.2) holds. □

REMARKS. (a) It would be fair to say that one can almost always make (3.1) hold by replacing X by a suitable equivalent process; see e.g. [6], p. 23; see also [5].

(b) Each set belonging to $\mathcal{O}$ is finely open. However, $\mathcal{O}$ can be much smaller than the collection of all finely open sets. (For Brownian motion in $\mathbb{R}^d$, $\mathcal{O}$ is the ordinary topology.)

(c) Suppose the hypotheses of Theorem 1.1 hold. Let μ be a probability measure on $\mathcal{E}^*$ and let $P = P^{\mu}$. Let $\mathcal{B}$ be the P-completion of $\mathcal{M}$ and let $\mathcal{B}_t^o$ be the P-completion of $\mathcal{M}_t$ in $\mathcal{B}$. Let $\mathcal{E}^e$ be as in Theorem 1.1. For $f \in \mathcal{E}_+^e$, the process $(f(X_t))$ is measurable so we can define $U^\alpha f(x) = E^x[\int_0^\infty e^{-\alpha t} f(X_t)dt]$ for $0 < \alpha < \infty$ and for $x \in E$. Then $(U^\alpha)_{0 < \alpha < \infty}$ is a Markov resolvent on $(E, \mathcal{E}^e)$ whose collection of β-excessive functions is precisely S^β for $\beta \geq 0$. It follows ([8], p. 242, T64) that (3.0) holds. By (a) of Theorem 1.1, (3.1) holds. That (3.2) holds may be proved as in [9], p. 36. Finally, it follows from (a) and (b) of Theorem 1.1 that (3.3) holds, and it is clear that $\mathcal{O} \subseteq \mathcal{O}_f$. In this way, Theorem 3.1 implies Theorem 1.1.

REFERENCES

1. R.M. BLUMENTHAL and R.K. GETOOR. Standard processes and Hunt processes. *Proc. Symp. Markov Processes and Potential Theory*, 1967, 13-22. Wiley, New York, 1967.

2. R.M. BLUMENTHAL and R.K. GETOOR. *Markov Processes and Potential Theory*. Academic Press, New York, 1968.

3. D.W. BRESSLER and M. SION. The current theory of analytic sets. *Canad. J. Math. 16* (1964), 207-230.

4. C. DELLACHERIE and P.-A. MEYER. *Probabilités et potentiel*, Chapitres I a IV, édition entièrement refondue, Hermann, Paris,1975.

5. E.B. DYNKIN. Regular Markov processes. *Russian Math. Surveys 28:2* (1973), 33-64.

6. R.K. GETOOR. *Markov Processes: Ray Processes and Right Processes*. Lecture Notes in Math. *440*, Springer-Verlag, Berlin, 1975.

7. G.A. HUNT. Markov processes and potentials, I. *Illinois J. Math. 1* (1957), 44-93.

8. P.-A. MEYER. *Probabilités et potentiel*. Hermann, Paris, 1966.

9. P.-A. MEYER. *Processus de Markov*. Lecture Notes in Math. *26*, Springer-Verlag, Berlin, 1967.

10. P.-A. MEYER. Balayage pour les processus de Markov continus à droite, d'après Shih Chung Tuo. *Séminaire de Probabilités V*, pp. 270-274. Lecture Notes in Math. *191*, Springer-Verlag, Berlin, 1971.

11. P.-A. MEYER. Convergence faible et compacite des temps d'arrêt, d'après Baxter et Chacon. *Séminaire de Probabilités XII*, pp. 411-423. Lecture Notes in Math. *649*, Springer-Verlag, Berlin, 1978.

12. C.T. SHIH. On extending potential theory to all strong Markov processes. *Ann. Inst. Fourier 20, 1* (1970), 303-315.

13. J.B. WALSH and P.-A. MEYER. Quelques applications des résolvantes de Ray. *Invent. Math. 14* (1971), 143-166.

Department of Mathematics
The Ohio State University
Columbus, Ohio 43210

Seminar on Stochastic Processes, 1983
Birkhäuser, Boston, 1984

CAPACITY THEORY AND WEAK DUALITY

by

R. K. GETOOR*

1. Introduction

In [8] Hunt developed his celebrated capacity theory for Markov processes. He assumed the underlying process satisfied his hypotheses (F) and (G). Hypothesis (F) is essentially a duality hypothesis together with a strong Feller condition, and hypothesis (G) is a transience condition. Portions of this theory are presented in section VI-4 of [1]. Since then there have been many variations on this theme. Of particular note are [2], [12], and [13].

One of the nice things about Hunt's theory is that it enables one to extend Spitzer's asymptotic formula [14]. Spitzer showed that if X is Brownian motion in $\mathbb{R}^d$ with $d \geq 3$ and B is a compact subset of $\mathbb{R}^d$, then

$$\int dx\, P^x(T_B \leq t) \sim t\, C(B) \tag{1.1}$$

as $t \to \infty$. Here $T_B = \inf\{t > 0: X_t \in B\}$ is the hitting time of B, dx is Lebesgue measure in $\mathbb{R}^d$, and C(B) is the Newtonian capacity of B properly normalized. Shortly thereafter in [3], I extended (1.1) to the case in which X satisfies (F) and (G) and the dual $\hat{X}$ of X has an in-

*This research was supported, in part, by NSF Grant MCS 79-23922.

finite lifetime. Actually the proof of (1.1) and its extension are very easy. The main content of [14] and [3] is to obtain the second term in the asymptotic expansion of (1.1) for various special processes. It also is quite easy to extend (1.1) to the case in which it is not assumed that $\hat{X}$ has an infinite lifetime. One then must replace C(B) by $\int \hat{P}^x(\zeta = \infty)\pi_B(dx)$ where π_B is the capacitary measure of B. See (2.16) for the precise statement.

An appropriate setting in which to study capacity and related results such as (1.1) is a pair of Borel right processes X and $\hat{X}$ in weak duality as discussed in [7]. This is general enough to cover all previous developments known to me. In this generality it is necessary to distinguish between the capacitary measure π_B and the left capacitary measure π_{B-} of a set B. Such a theory was begun in section 13 of [7], but the condition of strong transience assumed there is too restrictive. This is discussed in section 8 of the present paper. See also section 2.

In section 2 we describe in detail the main results of this paper. It contains definitions, statements of results, and discussion. The reader may want to skip to that section and return to this introduction only as needed. Sections 3 through 8 contain proofs and further discussion. Some elementary examples are given in section 9 to illustrate the necessity of certain hypotheses and the limitations of some of the results. The remainder of this introduction is devoted to setting out the precise hypotheses that will be in force throughout this paper and to recalling some facts from [5] and [7] that will be used frequently in the sequel.

Let $X = (\Omega, \mathcal{F}, \mathcal{F}_t, X_t, \theta_t, P^x)$ and $\hat{X} = (\hat{\Omega}, \hat{\mathcal{F}}, \hat{\mathcal{F}}_t, \hat{X}_t, \ddot{\theta}_t, \hat{P}^x)$ be Borel right processes on a Lusin topological space E (that is, E is homeomorphic to a Borel subset of a compact metric space) with Borel σ-algebra $\mathcal{E}$. A point $\Delta \notin E$ will serve as cemetery point. A Borel right process on a

Lusin state space is nothing but a right continuous, strong Markov process without branch points and having a Borel measurable semigroup. Let P_t (resp. $\hat{P}_t$) and U^α (resp. $\hat{U}^\alpha$) denote the semigroup and resolvent of X (resp. $\hat{X}$). Let m be a fixed σ-finite measure on E. If

$$(1.2) \qquad \int P_t f \cdot g \, dm = \int f \cdot \hat{P}_t g \, dm$$

for $t \geq 0$ and all positive Borel functions f and g on E, and if

$$(1.3) \qquad \begin{array}{l} X_{t-} \text{ exist in E for all } t \in \,]0,\zeta[\\ \hat{X}_{t-} \text{ exist in E for all } t \in \,]0,\hat{\zeta}[, \end{array}$$

then X and $\hat{X}$ are said to be in *weak duality* with respect to m (or the triple $(X,\hat{X},m)$ is in weak duality). If, in addition, for every $\alpha \geq 0$ and $x \in E$

$$(1.4) \qquad U^\alpha(x,\cdot) << m \quad \text{and} \quad \hat{U}^\alpha(x,\cdot) << m,$$

then X and $\hat{X}$ are in *strong duality* with respect to m ($(X,\hat{X},m)$ is in strong duality). In [15], Walsh showed that under (1.2), (1.3) holds almost surely P^x for m almost all x (i.e. almost surely P^m), and it follows that under (1.2) and (1.4), (1.3) holds almost surely (i.e. almost surely P^x for all x). If $(X,\hat{X},m)$ is in strong duality, then for each $\alpha \geq 0$ a potential density $u^\alpha(x,y)$ can be chosen which is $\mathcal{E} \times \mathcal{E}$ measurable and so that

$$(1.5) \qquad U^\alpha(x,dy) = u^\alpha(x,y)m(dy), \quad \hat{U}^\alpha(x,dy) = u^\alpha(y,x)m(dy),$$

and $x \to u^\alpha(x,y)$ is α-excessive (for X) for each y, and $y \to u^\alpha(x,y)$ is α-coexcessive (α-excessive for $\hat{X}$) for each x. A systematic study of weak duality may be found in [7]. We shall say that X is *transient* if there exists a strictly positive Borel function q such that Uq is bounded. See [4] for some apparently weaker conditions equivalent to this. Of course, transience of $\hat{X}$ is defined similarly.

It will be convenient to assume that X and $\hat{X}$ are both defined on the canonical path space. In other words let Ω be the space of all right continuous maps $\omega: \mathbb{R}^+ \to E \cup \{\Delta\}$ with Δ as cemetery and having left limits $\omega(t-)$ in E for $0 < t < \zeta(\omega) = \inf\{t: \omega(t) = \Delta\}$. Then both X_t and $\hat{X}_t$ are the coordinate maps on Ω, $X_t(\omega) = \omega(t) = \hat{X}_t(\omega)$ and the processes are completely described by the families of measures $(P^x: x \in E)$ and $(\hat{P}^x: x \in E)$ on $(\Omega, \mathcal{F}^0)$ where $\mathcal{F}^0$ is the σ-algebra generated by the coordinate maps. We shall assume that X and $\hat{X}$ are so defined throughout this paper.

In strong duality, one can use the potential densities $u^\alpha(x,y)$ to define the potential of a positive measure μ by

$$(1.6) \qquad U^\alpha\mu(x) = \int u^\alpha(x,y)\mu(dy).$$

By Fubini's theorem $\mu\hat{U}^\alpha(dx) = U^\alpha\mu(x)m(dx)$. It is this relationship that is used in defining the "potential of a measure" in weak duality, where potential densities may not exist. Let $\mathcal{S}^\alpha$ be the class of *Borel measurable* α-excessive functions which are finite a.e. m. It is shown in [7, (6.11)] that if f is α-excessive then there exists a Borel measurable α-excessive function g with $f = g$ a.e. m. A function $u \in \mathcal{S}^\alpha$ will be called the α-*potential function* of a measure μ provided $\mu\hat{U}^\alpha(dx) = u(x)m(dx)$. We write $u = U^\alpha(\mu)$. Clearly $\mu\hat{U}^\alpha$ is σ-finite since u is finite m a.e. and the transience of $\hat{X}$ implies that μ itself is σ-finite even when $\alpha = 0$. Let $\mathfrak{M}^\alpha$ be the class of measures μ such that $\mu\hat{U}^\alpha$ is σ-finite and absolutely continuous with respect to m. It is shown in [5] that for such a μ one may choose a $u \in \mathcal{S}^\alpha$ so that $\mu\hat{U}^\alpha = u\,m$. Here $u\,m$ is the measure $u(x)m(dx)$. Thus $u = U^\alpha(\mu)$. Of course, u is only determined a.e. m by μ, and so all identities between potential functions must be interpreted as holding a.e. m. In [5], it is shown that a measure μ is in $\mathfrak{M}^\alpha$ if and only if $\mu\hat{U}^\alpha$ is σ-finite and μ does not charge cofinely open m-copolar sets. (A Borel set B is m-copolar provided $\hat{P}^m(T_B < \infty) = 0$. See [7,§6] for a discussion of such things.)

As usual when $\alpha = 0$ we drop it from our notation. Thus if $\mu \in \mathfrak{M} = \mathfrak{M}^0$, $U(\mu)$ is its potential function so that $\mu\hat{U} = U(\mu)m$.

If B is a Borel set, then

$$(1.7) \quad T_B = \inf\{t > 0: X_t \in B\}, \quad S_B = \inf\{0 < t < \zeta; X_{t-} \in B\}$$

are the *hitting time* and *left hitting time* of B respectively. Here the infimum of the empty set is $+\infty$. Both T_B and S_B are perfect terminal times and we define

$$(1.8) \quad P_B^\alpha f(x) = E^x[e^{-\alpha T_B} f(X_{T_B})], \quad P_{B-}^\alpha f(x) = E^x[e^{-\alpha S_B} f(X_{S_B})].$$

Of course, $\hat{P}_B^\alpha$ and $\hat{P}_{B-}^\alpha$ are defined similarly relative to $\hat{X}$. A set $B \in \mathcal{E}$ is m-*polar* (resp. *left* m-*polar*) provided $P^m(T_B < \infty) = 0$ (resp. $P^m(S_B < \infty) = 0$). We now list some properties of potential functions that will be used in the sequel. They are all proved in [5].

(1.9) If $\mu, \nu \in \mathfrak{M}^\alpha$ and $U^\alpha(\mu) = U^\alpha(\nu)$ a.e. m, then $\mu = \nu$.

(1.10) If $u \in \mathcal{S}^\alpha$ and $u \le U^\alpha(\nu)$ a.e. m where $\nu \in \mathfrak{M}^\alpha$, then there exists $\mu \in \mathfrak{M}^\alpha$ with $u = U^\alpha(\mu)$.

(1.11) If $\mu \in \mathfrak{M}^\alpha$ and $\hat{\mu} \in \hat{\mathfrak{M}}^\alpha$, then for each $\beta \ge \alpha$, $\int U^\beta(\mu)d\hat{\mu} = \int \hat{U}^\beta(\hat{\mu})d\mu$.

(1.12) Let $\mu \in \mathfrak{M}^\alpha$ and $B \in \mathcal{E}$. Then $\mu\hat{P}_B^\alpha$ and $\mu\hat{P}_{B-}^\alpha$ are in $\mathfrak{M}^\alpha$ and their α-potential functions are $P_{B-}^\alpha U^\alpha(\mu)$ and $P_B^\alpha U^\alpha(\mu)$ respectively.

The duals of (1.9), (1.10) and (1.12) are equally valid.

We introduce the notation (f,g) for $\int fg\,dm$ whenever it exists. Thus (1.2) may be written $(P_t f,g) = (f,\hat{P}_t g)$. It follows that $(U^\alpha f,g) = (f,\hat{U}^\alpha g)$ for positive Borel functions f and g. If $B \in \mathcal{E}$ the following identity for such f and g is proved in [7,(11.3)]:

$$(1.13) \qquad (P_B^\alpha U^\alpha f,g) = (f,\hat{P}_{B-}^\alpha \hat{U}^\alpha g).$$

It is the weak duality version of Hunt's switching identity. If both X and $\hat{X}$ are standard processes, then it is shown in [7,(15.2) and (15.7)] that almost surely $T_B \le S_B$ and $P^m(T_B \ne S_B) = 0$. It follows that $P^x(T_B \ne S_B) = 0$ except possibly for x in a finely open m-polar set.

Notation. We use the symbol "$\equiv$" to mean "is defined to be." If $(W,\mathcal{W})$ is a measurable space, $\mathcal{W}$ (resp. $\mathcal{W}^+$, $b\mathcal{W}$, $b\mathcal{W}^+$) will also denote the collection of $\mathcal{W}$ measurable numerical functions on W (resp. which are positive, which are bounded, which are positive and bounded). A measure is always a positive measure unless explicitly stated otherwise. If $f \in \mathcal{E}^+$ and ν is a measure on E, then $f\nu$ denotes the measure $f(x)\nu(dx)$ while $\nu f = \nu(f) = \int f d\nu$. In more complicated formulas we shall sometimes write $f\cdot\nu$ or $\nu\cdot f$ for clarity in place of $f\nu$. For example, $\mu\hat{U}\cdot g$ is the measure $A \to \int_A \mu\hat{U}(dx)g(x)$ while $\mu\hat{U}g = \int \mu\hat{U}(dx)g(x)$.

2. Description of Results

Throughout the remainder of this paper $(X,\hat{X},m)$ is in weak duality as described in section 1, and it is assumed that both X and $\hat{X}$ are *transient*. We introduce the following notation. If $B \in \mathcal{E}$,

$$(2.1) \qquad \phi_B(x) = P^x(T_B < \infty), \quad \psi_B(x) = P^x(S_B < \infty)$$

where T_B and S_B are the hitting time and left hitting time of B defined in (1.7). Of course, $\hat{\phi}_B$ and $\hat{\psi}_B$ are defined similarly relative to $\hat{X}$. We also define

(2.2) $$L_B = \sup\{t\colon X_t \in B\},\quad M_B = \sup\{0 < t < \zeta\colon X_{t-} \in B\}$$

and similarly $\hat{L}_B$ and $\hat{M}_B$. In these definitions the supremum of the empty set is taken to be zero. Since $B \subset E$, $L_B \le \zeta$ and $M_B \le \zeta$. Observe that

(2.3) $$\phi_B(x) = P^x(L_B > 0),\quad \psi_B(x) = P^x(M_B > 0).$$

(2.4) DEFINITION. *Let* $B \in \mathbf{E}$. *Then,* B *is* m-*transient (resp. left* m-*transient) provided there exists a measure* π_B *(resp.* π_{B-}*) on* E *such that* $\phi_B = U(\pi_B)$ *(resp.* $\psi_B = U(\pi_{B-})$*).*

Remember that $U(\pi_B)$ denotes an element of $\mathcal{S}$, say u, such that $\pi_B\hat{U} = u\,m$. In particular, $U(\pi_B)$ is Borel measurable so one should really write $\phi_B = U(\pi_B)$ a.e. m in (2.4). However, our convention is that equalities or inequalities involving potential functions are to be interpreted as holding a.e. m. Similar comments apply to $\psi_B = U(\pi_{B-})$ in (2.4).

We call π_B the *capacitary* measure of B and π_{B-} the *left capacitary* measure of B. If $q > 0$ is such that $m(q) < \infty$ and $h \equiv \hat{U}q \le 1$ (possible since m is σ-finite and $\hat{X}$ is transient), then

$$\pi_B(h) = \pi_B\hat{U}(q) = \int \phi_B\, q\, dm < \infty,$$

and since $h > 0$, π_B and $\pi_B\hat{U}$ are σ-finite. Similarly π_{B-} and $\pi_{B-}\hat{U}$ are σ-finite. It now follows (1.9) that π_B and π_{B-} are uniquely determined provided they exist. The dual objects $\hat{\pi}_B$ and $\hat{\pi}_{B-}$ are defined if B is m-cotransient (i.e. m-transient for $\hat{X}$) or left m-cotransient.

(2.5) DEFINITION. *If* B *is* m-*transient,* $C(B) = \pi_B(1)$ *is the capacity of* B. $C_-(B) = \pi_{B-}(1)$ *is the left capacity of* B *if* B *is left* m-*transient.*

The *cocapacity* of B, $\hat{C}(B) = \hat{\pi}_B(1)$ and *left cocapacity* $\hat{C}_-(B) = \hat{\pi}_{B-}(1)$ are defined whenever B satisfies the appropriate cotransience condition.

(2.6) REMARK. Since m is fixed in this discussion we will say B is transient rather than m-transient when no confusion is possible. A similar convention holds for the other types of transience. Obviously, if B is transient, then $C(B) = 0$ if and only if $\phi_B = P_B 1 = 0$ a.e. m; that is, if and only if B is m-polar. Similarly for a left transient set B, $C_-(B) = 0$ if and only if B is left m-polar.

(2.7) REMARK. If both X and $\hat{X}$ are standard, then as stated in section 1, $T_B = S_B$ a.s. P^m and so $\phi_B = \psi_B$ a.e. m. But this implies that B is transient if and only if it is left transient and that for such B, $\pi_B = \pi_{B-}$ according to (1.9). In particular $C(B) = C_-(B)$ in this case.

(2.8) PROPOSITION. *If* B *is left transient,* π_{B-} *is carried by* $\bar{B}$.

If B is transient, π_B need not be carried by $\bar{B}$. See example (9.2).

We now give some simple sufficient conditions that cover the "classical" situation.

(2.9) PROPOSITION. *Let* $\mu \in \mathfrak{M}$ *and suppose* $B \subset \{U(\mu) \geq 1\}$ *up to an* m-*polar set. Then* B *is transient. If* $\bar{B} \subset \{U(\mu) \geq 1\}$ *up to a left* m-*polar set, then* B *is left transient. If* B *is left transient and* $\bar{B} \subset \{\hat{U}(\hat{\mu}) \geq 1\}$ *where* $\hat{\mu} \in \hat{\mathfrak{M}}$, *then* $C_-(B) \leq \int \psi_B d\hat{\mu}$. *In particular, if* $\hat{\mu}$ *is finite,* $C_-(B) < \infty$.

(2.10) REMARKS: THE CLASSICAL SITUATION. The following assumptions will be called the *classical situation* in the sequel. E is locally compact with a countable base and m is a Radon measure. The processes X and $\hat{X}$ are standard, and the α-excessive and α-coexcessive functions are lower semi-continuous (lsc) for some $\alpha > 0$. It is known that the transience conditions take the form $x \to U(x,K)$ and $x \to \hat{U}(x,K)$ are bounded for compact K under these conditions. See [4]. If B has compact closure $\bar{B}$ and G is a compact neighborhood of $\bar{B}$, then $U(x,G)$ and $\hat{U}(x,G)$ are bounded away from zero on $\bar{B}$, say by $a > 0$. Then $\bar{B} \subset \{U1_G \wedge \hat{U}1_G \geq a\}$ and so by (2.9), B is transient and $C(B) = C_-(B) < \infty$. Note $\hat{U}1_G = \hat{U}(1_G \cdot m)$ and $m(G) < \infty$.

The next result is one of the key facts in capacity theory. The proof we give in section 3 is much simpler than proofs known to me even under much stronger hypotheses. See, for example, [8],[1],[12], or [7].

(2.11) THEOREM. *If* B *is transient and left cotransient, then* $C(B) = \hat{C}_-(B)$. *If* B *is left transient and cotransient, then* $C_-(B) = \hat{C}(B)$.

If both X and $\hat{X}$ are standard, then for B transient and cotransient one has $C(B) = C_-(B) = \hat{C}(B) = \hat{C}_-(B)$. Here is a refinement of (2.8) in this situation.

(2.12) PROPOSITION. *Suppose* X *and* $\hat{X}$ *are standard and* B *is transient. Then* $\pi_{B-} = \pi_B$ *is carried by* $B \cup B^{cr}$ *where* B^{cr} *denotes the set of coregular points for* B.

Most likely π_{B-} is carried by $B \cup B^{cr}$ without the standardness assumption, but I have not succeeded in proving it. However, if $\bar{B} \subset \{U(\mu) = 1\}$ up to a left m-polar set for some $\mu \in \mathfrak{M}$, in particular if $\bar{B} \subset G$ with G open and left transient, then π_{B-} is carried by $B \cup B^{cr}$. See (3.5).

It is necessary to consider α-capacity for $\alpha > 0$. Define for $B \in \mathbf{E}$

(2.13) $$\phi_B^\alpha(x) = E^x(e^{-\alpha T_B}),\quad \psi_B^\alpha(x) = E^x(e^{-\alpha S_B}),$$

where $\phi_B^0 = \phi_B$ and $\psi_B^0 = \psi_B$ defined in (2.1). Then B is α-transient (resp. left α-transient) provided there exists a (necessarily σ-finite) measure π_B^α (resp. π_{B-}^α) such that $\phi_B^\alpha = U^\alpha(\pi_B^\alpha)$ (resp. $\psi_B^\alpha = U^\alpha(\pi_{B-}^\alpha)$). One defines $C^\alpha(B) = \pi_B^\alpha(1)$, $C_-^\alpha(B) = \pi_{B-}^\alpha(1)$. The corresponding dual objects relative to $\hat{X}$ are defined similarly. All of the results for $\alpha = 0$ extend in an obvious manner to $\alpha > 0$. For example, $C^\alpha(B) = \hat{C}_-^\alpha(B)$ and $C_-^\alpha(B) = \hat{C}^\alpha(B)$ whenever they exist, and if X and $\hat{X}$ are standard then $C^\alpha(B) = C_-^\alpha(B)$ and $\hat{C}^\alpha(B) = \hat{C}_-^\alpha(B)$.

(2.14) PROPOSITION. *Fix $\beta \geq 0$. If $B \in \mathcal{E}$ is β-transient (resp. left β-transient), then it is α-transient (resp. left α-transient) for every $\alpha > \beta$ and one has*

(2.15) (i) $$\pi_B^\alpha = \pi_B^\beta + (\alpha - \beta)(\phi_B^\beta\, m)\hat{P}_{B-}^\alpha$$

(ii) $$\pi_{B-}^\alpha = \pi_{B-}^\beta + (\alpha - \beta)(\psi_B^\beta\, m)\hat{P}_B^\alpha.$$

If $C^\alpha(B) < \infty$ for some $\alpha > \beta$, then π_B^α decreases setwise to π_B^β as α decreases to β and the analogous statement holds for the left capacitary measures.

Example (9.1) shows that the limit relations may be false without the condition $C^\alpha(B) < \infty$ for some $\alpha > \beta$ even if $C^\beta(B) < \infty$.

The following is the generalization of Spitzer's result mentioned in section 1.

(2.16) THEOREM. *Suppose B is transient and $C^\alpha(B) < \infty$ for some $\alpha > 0$. Then as $t \to \infty$,*

$$t^{-1} \int m(dx)P^x(T_B \leq t) \to \int \hat{P}^x(\zeta = \infty)\pi_B(dx).$$

The analogous result for S_B (with π_B replaced by π_{B-}) is equally valid. Let

$$\gamma(B) \equiv \int \hat{P}^x(\zeta = \infty)\pi_B(dx) = C(B) - \int \hat{P}^x(\zeta < \infty)\pi_B(dx)$$

denote the constant in (2.16). It will be shown in section 7 that $\gamma(B)$ is the "conditional" capacity of B when $\hat{X}$ is conditioned to have infinite lifetime. Moreover under the conditions of (2.16), one has

$$(2.17) \qquad t^{-1}\int m(dx)\hat{P}^x(\zeta = \infty)P^x(T_B \le t) \to \gamma(B) \quad \text{as } t \to \infty,$$

and the analogous statement for S_B is also valid. Example (9.1) shows that (2.16) and (2.17) may fail if one merely assumes $C(B) < \infty$.

In [7] a set $B \in \mathcal{E}$ is called *strongly* m-*transient* provided (recall $L_B \le \zeta$)

$$(2.18) \qquad P^m(L_B = \zeta) = 0.$$

In keeping with our convention for transience we shall call such a set *strongly transient* in this paper. Similarly B is strongly left transient provided $P^m(M_B = \zeta) = 0$. It was shown in [7] that if B is strongly transient then $\phi_B = U(\pi_B)$ where π_B is the Revuz measure of the (raw) additive functional

$$(2.19) \qquad K_t^B \equiv 1_{[L_B,\infty[}(t)\, 1_{\{L_B > 0\}}.$$

Thus in the terminology of this paper a strongly transient set B is transient and its capacitary measure π_B (using the definition of Revuz measure — see [7]) is given by

$$(2.20) \qquad \pi_B(f) = \lim_{t \to 0} t^{-1}E^m[f(X_{L_B-}); 0 < L_B \le t].$$

Similarly if B is strongly left transient, it is left transient and

$$(2.21)\qquad \pi_{B-}(f) = \lim_{t\to 0} t^{-1} E^m[f(X_{M_{B-}}); 0 < M_B \le t]$$

is the Revuz measure of $K_t^{B-} \equiv 1_{[M_B,\infty[}(t)\, 1_{\{M_B > 0\}}$.

Using the results in the appendix of [5], these interpretations of the capacitary measures may be extended to a wider class of sets. Fix $q \in \mathbf{E}$ with $0 < q \le 1$ and satisfying $m(q) < \infty$ and $h \equiv \hat{U}q \le 1$. Fix a countable set (g_n) which is uniformly dense in the bounded, positive, uniformly continuous functions on E and suppose $g_1 = 1$. Define

$$(2.22)\qquad \Lambda = \bigcap_r \bigcap_n \{0 < \zeta < \infty,\ X_{\zeta-} \text{ exists in E},$$
$$h(X_{\zeta-}) = h(X_\zeta)_-,\ \hat{U}^r(hg_n)(X_{\zeta-}) = \hat{U}^r(hg_n)(X_\zeta)_-\}$$

where $r > 0$ runs over the positive rationals. Suppose for a moment that we are in the classical situation (2.10) and that $\hat{U}^r$ maps continuous functions with compact support, $\underset{\sim}{C}_K$, into continuous functions vanishing at infinity, $\underset{\sim}{C}_0$. Then one may choose $q \in \underset{\sim}{C}_0$ so that $h \in \underset{\sim}{C}_0$ and $(g_n)_{n\ge 2} \subset \underset{\sim}{C}_K$. In this situation $\Lambda = \{0 < \zeta < \infty, X_{\zeta-}$ exists in E$\}$ almost surely.

Returning to the general case here is the extension of (2.20) and (2.21).

(2.23) THEOREM. *Let* B $\in \mathbf{E}$ *and suppose that* $\{L_B = \zeta\} \subset \Lambda$ a.s. P^m. *Then* B *is transient and* π_B *is the Revuz measure of* K^B *defined in* (2.19). *Hence* π_B *is given by* (2.20). *If* $\{M_B = \zeta\} \subset \Lambda$ a.s. P^m, B *is left transient and* π_{B-} *is given by* (2.21).

(2.24) REMARKS. Since $P^m(\zeta = 0) = 0$, one may replace $\{L_B = \zeta\}$ by $\{0 < L_B = \zeta\}$ in (2.23). Also if $\{L_B = \zeta\} \subset \Lambda$ a.s. P^m, then $P^m(L_B = \infty)$

$= 0$. Clearly if B is strongly transient the condition in (2.23) holds. The difference is that K^B does not charge ζ if B is strongly transient, while the jump of K^B at ζ, $\Delta K_\zeta^B = 1_{\{0 < L_B = \zeta\}}$ is carried by Λ a.s. P^m under the assumptions of (2.23).

Some sufficient conditions for a set to be strongly transient are given at the end of section 8. See also [6]. However, there are interesting situations in which strong transience does not hold. For example if X is a stable process in $\mathbb{R}^d$ killed when it leaves the unit ball E, then a compact subset B of E is not strongly transient, but it does satisfy the condition of (2.23), $\{L_B = \zeta\} \subset \Lambda$.

Theorem 7.9 of [5] gives a characterization of transient sets in terms of h-transforms which is not too different from the sufficient condition of (2.23). Namely, B is transient if and only if

$$P^{x/\phi_B}(\Lambda) = 1 \text{ a.e. } m \text{ on } \{\phi_B > 0\}. \tag{2.25}$$

This will be discussed in more detail in section 7. But we remind the reader here that X killed at L_B has for its semigroup the ϕ_B transform of (P_t). See [10].

Under strong transience assumptions the various capacitary measures have a very simple interpretation in terms of the stationary process Z built over $(X,\hat{X},m)$. These were established in section 13 of [7], but will be recalled in section 8. In section 7 we discuss conditional capacities. Surprisingly, these were introduced by Hunt [8] under some additional hypotheses. However, it is not clear if Hunt realized that the general capacities which he defined in section 19 of [8] were, in fact, conditional capacities. Finally section 9 contains some elementary examples illustrating some of the technical points discussed earlier.

There are several additional asymptotic results which complement (2.16).

(2.26) THEOREM. *Let* B *satisfy the hypotheses of* (2.23) *and for* $f \in \mathcal{E}^+$ *define*

$$H_B(f,t) = \int m(dx)E^x[f(X_{L_B-}); 0 < L_B \le t].$$

Then $t^{-1}H_B(f,t) \to \pi_B(f)$ *as* $t \to 0$. *If* $\pi_B(f) < \infty$, *then*

$$t^{-1}H_B(f,t) \to \int f(x)\hat{P}^x(\zeta = \infty)\pi_B(dx)$$

as $t \to \infty$. *If, in addition,* $\hat{P}^x(\zeta = \infty) = 1$ a.e. π_B, *then* $H_B(f,t) = t\pi_B(f)$.

(2.27) THEOREM. *Let* B *be strongly left cotransient. If* $f \in \mathcal{E}^+$, *then*

$$\int m(dx)\hat{P}^x(S_B = \infty)E^x[f(X_{T_B}); 0 < T_B \le t] = t\hat{\pi}_{B-}(f).$$

The proofs of (2.16) and (2.26) are elementary while the proof of (2.27) uses the interpretation of $\hat{\pi}_{B-}$ in terms of the stationary process Z. However, the following result is elementary and should be compared with (2.26) and (2.27).

(2.28) THEOREM. *Let* B *satisfy the hypotheses of* (2.23) *and let* $f \in \mathcal{E}^+$. *Then*

$$\int m(dx)\hat{P}^x(\zeta = \infty)E^x[f(X_{L_B-}); 0 < L_B \le t] = t\int \hat{P}^x(\zeta = \infty)f(x)\pi_B(dx).$$

Of course, the versions of (2.26), (2.27), and (2.28) corresponding to the other capacitary measures are equally valid.

3. Proof of (2.8), (2.9), (2.11), and (2.12)

We begin by proving (2.8). We need the following lemma.

(3.1) LEMMA. *Let* $B \in \mathbf{E}$ *and* G *be an open set with* $B \subset G$. *Then* $S_G + S_B \circ \theta_{S_G} = S_B$ *and* $P_{B-} = P_{G-} P_{B-}$.

PROOF. The second assertion is an immediate consequence of the first and so we need show only $S_G + S_B \circ \theta_{S_G} = S_B$. Since S_B is a terminal time this is clear if $S_G < S_B$. But $S_G \le S_B$ and so the desired conclusion obtains if $S_G = \infty$ or $S_B = 0$. The remaining case is $0 < S_G = S_B < \infty$. In this situation $X_{t-} \in G^c$ for $t < S_G$ and so $X_{S_G -} \in G^c$. Thus $X_{S_B -} = X_{S_G -} \notin B$, and hence there exists a sequence (t_n) with $X_{t_n -} \in B$, $t_n > S_B$, and $t_n \downarrow S_B = S_G$. Consequently, $S_B \circ \theta_{S_G} = 0$. □

PROOF OF (2.8). Since $P_{B-}1 = \psi_B = U(\pi_{B-})$, (3.1) and (1.12) give

$$U(\pi_{B-}) = P_{G-}U(\pi_{B-}) = U(\pi_{B-}\hat{P}_G),$$

a.e. m, and so, by uniqueness, $\pi_{B-} = \pi_{B-}\hat{P}_G$. But $\hat{P}_G(x,\cdot)$ is carried by $\bar{G}$ for each $x \in E$, and choosing G_n open with $\bar{B} \subset G_n$ and $\bar{B} = \cap \bar{G}_n$ we see that π_{B-} is carried by $\bar{B}$. □

PROOF OF (2.9). Suppose $B \subset \{U(\mu) \ge 1\}$ up to an m-polar set. Since an m-polar set is contained in a Borel finely closed m-polar set [7, (6.12)], it follows that a.e. m one has

$$\phi_B = P_B 1 \le P_B U(\mu) = U(\mu\hat{P}_{B-}).$$

Thus ϕ_B is dominated m a.e. by the potential of a measure in $\mathfrak{M}$, and hence B is transient by (1.10). Next suppose $\bar{B} \subset \{U(\mu) \ge 1\}$ up to a left m-polar set. Since P_{B-} is carried by $\bar{B}$ one has $\psi_B = P_{B-}1 \le P_{B-}U(\mu) = U(\mu\hat{P}_B)$, and, as before, B is left transient. Finally, suppose B is left m-transient and $\bar{B} \subset \{\hat{U}(\hat{\mu}) \ge 1\}$ with $\hat{\mu} \in \hat{M}$. Then by (2.8)

$$C_-(B) = \pi_{B-}(\bar{B}) \le \int \hat{U}(\hat{\mu})d\pi_{B-} = \int U(\pi_{B-})d\hat{\mu} = \int \psi_B d\hat{\mu},$$

where the second equality uses (1.11), and the third the fact that $\{\psi_B \neq U(\pi_{B-})\}$ is finely open and m-null, hence m-polar, and $\hat{\mu} \in \hat{M}$ does not charge such sets. □

PROOF OF (2.11). Let B be transient and left cotransient. Choose sequences (f_n) and (g_k) of Borel functions such that $Uf_n \uparrow 1$ and $\hat{U}g_k \uparrow 1$. This is possible since X and $\hat{X}$ are transient. Then because of (1.13)

$$(3.2) \qquad A_{n,k} \equiv (P_B Uf_n, g_k) = (f_n, \hat{P}_{B-}\hat{U}g_k).$$

Consequently $A_{n,k}$ increases with both n and k. Moreover

$$\lim_n A_{n,k} = (P_B 1, g_k) = (U(\pi_B), g_k) = \int \hat{U}g_k d\pi_B \uparrow C(B) \quad \text{as } k \to \infty.$$

Similarly $\lim_n \lim_k A_{n,k} = \hat{C}_-(B)$. □

(3.3) REMARK. The above argument shows that if g is any coexcessive function and $\hat{U}g_k \uparrow g$, then

$$\int g d\pi_B = \lim_k (\phi_B, g_k) = \lim_k \lim_n (P_B Uf_n, g_k)$$

where $Uf_n \uparrow 1$. Moreover $(P_B Uf_n, g_k) = (f_n, \hat{P}_{B-}\hat{U}g_k)$ increases with both n and k. In particular if $\hat{U}g_k \uparrow 1$, then $C(B) = \lim_k (\phi_B, g_k)$. Since $(\phi_B, g_k) = \int \hat{U}g_k d\pi_B$ it increases with k. But ϕ_B increases with B and so C(B) is an increasing function of B. Moreover if $B_n \uparrow B$, then $T_{B_n} \downarrow T_B$ so that $\phi_{B_n} \uparrow \phi_B$, and consequently $C(B_n)$ increases to C(B), provided B is transient.

PROOF OF (2.12). This is just a re-working of the argument at the top of page 287 of [1] in the present situation. Since X and $\hat{X}$ are standard, $\phi_B = \psi_B$ a.e. m, and so $\pi_B = \pi_{B-}$. Let ν' be the restriction of

π_B to B and let $\nu = \pi_B - \nu'$. If G is open and $B \subset G$, then by (3.1)

$$\psi_B = P_{G^-}\psi_B = P_{G^-}U(\nu') + P_{G^-}U(\nu) = U(\nu'\hat{P}_G) + U(\nu\hat{P}_G).$$

But ν' is carried by $B \subset G$ and so $\nu'\hat{P}_G = \nu'$. Thus

$$U(\pi_B) = U(\nu') + U(\nu\hat{P}_G) \tag{3.4}$$

a.e. m for any open set $G \supset B$. Using the standardness of $\hat{X}$ and $\nu(B) = 0$ choose a decreasing sequence (G_n) of open sets containing B such that $T_{G_n} \wedge \zeta \uparrow T_B \wedge \zeta$ almost surely $\hat{P}^\nu$. Now $U(\nu\hat{P}_{G_n}) = P_{G_n^-}U(\nu)$ a.e. m, and this last expression decreases since $U(\nu)$ is excessive. Thus $U(\nu\hat{P}_{G_n})$ decreases to a Borel function, say g, a.e. m.

Next choose $f > 0$ with $\int f\,dm < \infty$. Then

$$\int f\,U(\nu\hat{P}_{G_n})dm \le \int f\,\phi_B dm < \infty,$$

and

$$(f, U(\nu\hat{P}_{G_n})) = \int \hat{U}f\,d(\nu\hat{P}_{G_n}) = \int \hat{P}_{G_n}\hat{U}f\,d\nu = \hat{E}^\nu \int_{T_{G_n}}^{\zeta} f(X_t)dt.$$

But as n approaches infinity this last expression approaches

$$\hat{E}^\nu \int_{T_B}^{\zeta} f(X_t)dt = \int \hat{P}_B\hat{U}f\,d\nu = \int \hat{U}f\,d(\nu\hat{P}_B) = \int f\,U(\nu\hat{P}_B)dm.$$

Thus $g = U(\nu\hat{P}_B)$ a.e. m. Combining this with (3.4) we see that

$$U(\pi_B) = U(\nu') + U(\nu\hat{P}_B) = U(\nu' + \nu\hat{P}_B)$$

a.e. m, and so $\pi_B = \nu' + \nu\hat{P}_B$. Since $\hat{P}_B$ is carried by $B \cup B^{cr}$ this establishes (2.12). □

(3.5) REMARK. If $\bar{B} \subset \{U(\mu) = 1\}$ up to a left m-polar set with $\mu \in \mathfrak{M}$, then $\psi_B = P_{B-}U(\mu) = U(\mu\hat{P}_B)$ a.e. m. Consequently $\pi_{B-} = \mu\hat{P}_B$ is carried by $B \cup B^{cr}$. If G is open and transient, then $P_{G-}1 = 1$ on G, and so if $\bar{B} \subset G$, $\pi_{B-} = \pi_{G-}\hat{P}_B$.

4. α-Capacities: Proof of (2.14)

Fix $B \in \mathcal{E}$. For typographical convenience let $\phi^\alpha = \phi^\alpha_B$ and $\psi^\alpha = \psi^\alpha_B$ for any $\alpha \ge 0$. See (2.13). Fix for the moment $0 \le \beta < \alpha$. Routine calculations show

$$(4.1)\quad \text{(i)}\quad \phi^\beta - \phi^\alpha = (\alpha - \beta)[U^\alpha\phi^\beta - P^\alpha_B\, U^\alpha\, \phi^\beta]$$

$$\text{(ii)}\quad \psi^\beta - \psi^\alpha = (\alpha - \beta)[U^\alpha\psi^\beta - P^\alpha_{B-}U^\alpha\, \psi^\beta].$$

If B is β-transient so that $\phi^\beta = U^\beta(\pi^\beta_B)$, then by the resolvent equation — more exactly [5, (6.1)] — one has a.e. m

$$(4.2)\qquad \phi^\beta = U^\alpha(\pi^\beta_B) + (\alpha - \beta)U^\alpha\phi^\beta.$$

Combining this with (4.1-i) we obtain

$$(4.3)\quad \phi^\alpha = U^\alpha(\pi^\beta_B) + (\alpha - \beta)P^\alpha_B U^\alpha\phi^\beta = U^\alpha(\pi^\beta_B) + (\alpha - \beta)U^\alpha((\phi^\beta m)\hat{P}^\alpha_{B-}).$$

Defining

$$(4.4)\qquad \pi^\alpha_B = \pi^\beta_B + (\alpha - \beta)(\phi^\beta m)\hat{P}^\alpha_{B-},$$

we see that $U^\alpha(\pi^\alpha_B) = \phi^\alpha$. Hence B is α-transient and its α-capacitary measure is given by (4.4). Similarly if B is β left-transient it is α left-transient and π^α_{B-} is given by (2.15-ii).

It is obvious from (4.4) that $\alpha \to \pi_B^\alpha$ increases setwise. Let $\nu^\beta(A) = \lim_{\alpha \downarrow \beta} \pi_B^\alpha(A)$. If $C^\gamma(B) = \pi_B^\gamma(1) < \infty$ for some $\gamma > \beta$, then ν^β is a finite measure. By the resolvent equation

$$(4.5) \qquad U^\beta(\pi_B^\alpha) = U^\alpha(\pi_B^\alpha) + (\alpha - \beta)U^\beta U^\alpha(\pi_B^\alpha) = \phi^\alpha + (\alpha - \beta)U^\beta\phi^\alpha$$

a.e. m. Now $\phi^\alpha \uparrow \phi^\beta$ as $\alpha \downarrow \beta$ and if $\beta > 0$, $(\alpha - \beta)U^\beta\phi^\alpha \le (\alpha - \beta)U^\beta 1 \to 0$ as $\alpha \downarrow \beta$. Thus the right hand side of (4.5) approaches $\phi^\beta = U^\beta(\pi_B^\beta)$ as $\alpha \downarrow \beta$ at least when $\beta > 0$. If $f \in b\mathcal{E}^+$ and $\hat{U}^\beta f$ is bounded, then

$$(f, U^\beta(\pi_B^\alpha)) = \int \hat{U}^\beta f \, d\pi_B^\alpha \downarrow \int \hat{U}^\beta f \, d\nu^\beta = (f, U^\beta(\nu^\beta)),$$

as $\alpha \downarrow \beta$. Consequently $U^\beta(\nu^\beta) = U^\beta(\pi_B^\beta)$ a.e. m and so $\pi_B^\beta = \nu^\beta$. Thus π_B^α decreases to π_B^β setwise as α decreases to $\beta > 0$.

The only place the above argument breaks down when $\beta = 0$ is in showing that $\alpha U \phi^\alpha \to 0$ as $\alpha \to 0$. The proof of this in [8] or [12] makes essential use of weak convergence of measures. We shall give a proof valid under our present assumptions. From (4.5) when $\beta = 0$, $0 < \alpha < \gamma$, $\alpha U \phi^\alpha \le U(\pi_B^\alpha) \le U(\pi_B^\gamma)$ a.e. m. Let $q > 0$ with $\hat{U}q \le 1$. Then $(q, U(\pi_B^\gamma)) = \int \hat{U} q \, d\pi_B^\gamma \le C^\gamma(B) < \infty$. Thus $U(\pi_B^\gamma)$ is finite a.e. m and hence so is $\alpha U \phi^\alpha$ for $\alpha < \gamma$. In order to apply the argument of the preceding paragraph it suffices to show that $\alpha U \phi^\alpha \to 0$ a.e. m as α decreases to zero through some fixed sequence $A = (\alpha_n)$. Now the set of x such that for each $\alpha \in A$, $\alpha U \phi^\alpha(x) \le U(\pi_B^\gamma)(x) < \infty$ has full m measure. Given such an x,

$$\alpha U \phi^\alpha(x) = \alpha E^x \int_0^\infty e^{-\alpha T_B \circ \theta_t} \, dt.$$

Now $T_B \circ \theta_t = T_B - t$ if $t < T_B$ and $T_B \circ \theta_t = \infty$ if $t \ge L_B$. Therefore

$$(4.6) \qquad \alpha U \phi^\alpha(x) = \alpha E^x \int_0^{T_B} e^{-\alpha(T_B - t)} dt + \alpha E^x \int_{T_B}^{L_B} e^{-\alpha T_B \circ \theta_t} \, dt.$$

The first term on the right side of (4.6) equals

$$\alpha E^x\Big[\int_0^{T_B} e^{-\alpha t}dt;\ T_B < \infty\Big] = E^x\big[1 - e^{-\alpha T_B};\ T_B < \infty\big] \to 0 \quad \text{as } \alpha \to 0.$$

In order to handle the second term consider the closed homogeneous random set M in $]0,\infty[$ which is the closure in $]0,\infty[$ of $\{t + T_B \circ \theta_t;\ t > 0\}$. In the present case M is the closure in $]0,\infty[$ of $\{t > 0: X_t \in B\}$. Clearly $M \subset [T_B, L_B] \cap\ ]0,\infty[$. Let $\{(\ell_\nu, r_\nu),\ \nu \geq 0\}$ be the contiguous intervals to M in $]0,\infty[$ with $(r_0, \ell_0) =]0, T_B[$ if $T_B > 0$. Now $X_t \in B^r$ — the regular points for B — if $T_B \circ \theta_t = 0$, while if $\ell_\nu \leq t < r_\nu < \infty$ then $T_B \circ \theta_t = r_\nu - t$. Therefore the second term on the right side of (4.6) may be written

$$\alpha E^x \int_0^\infty 1_{B^r}(X_t)dt + \alpha E^x\Big\{\sum_{\nu\geq 1} \int_{\ell_\nu}^{r_\nu} e^{-\alpha(r_\nu - t)}dt;\ L_B > 0\Big\}.$$

Since this is finite, the first piece goes to zero as $\alpha \to 0$. The second piece equals

$$E^x\Big\{\sum_{\nu\geq 1}(1 - e^{-\alpha(r_\nu - \ell_\nu)});\ r_\nu < \infty,\ L_B > 0\Big\}. \tag{4.7}$$

Now $1 - e^{-\alpha(r_\nu - \ell_\nu)} \to 0$ as $\alpha \to 0$ if $r_\nu < \infty$ and since the expression in (4.7) is finite for $\alpha \leq \gamma$, it follows from the dominated convergence theorem that it approaches zero as $\alpha \to 0$. This completes the proof that $\pi_B^\alpha \downarrow \pi_B^\beta$ as $\alpha \downarrow \beta$ in all cases provided $C^\gamma(B) < \infty$ for some $\gamma > \beta$.

The argument for the left capacitary measures is exactly the same except that in the last part one uses the closure in $]0,\infty[$ of $\{t + S_B \circ \theta_t;\ t > 0\}$ in place of M and $F = \{x: P^x(S_B = 0) = 1\}$ in place of B^r.

5. Asymptotics I: Proof of (2.16) and (2.17)

Suppose $B \in \mathcal{E}$ is transient and define

(5.1) $$e_B(t) = \int m(dx)P^x(T_B \le t).$$

Since B is transient it is α-transient for each $\alpha > 0$ and $\phi^\alpha \equiv \phi^\alpha_B = U^\alpha(\pi^\alpha_B)$. Then

(5.2) $$E_B(\alpha) \equiv \int m(dx)\phi^\alpha(x) = \int \hat{U}^\alpha 1\, d\pi^\alpha_B.$$

If $C^\beta(B) < \infty$ for some $\beta > 0$, then $E_B(\beta) \le \beta^{-1}C^\beta(B) < \infty$. Since $P^x(T_B \le t) \le e^{\beta t}\phi^\beta(x)$, it follows that $e_B(t) < \infty$ if $t < \infty$. But then $\int_0^\infty e^{-\alpha t} de_B(t) = E_B(\alpha)$ and since this is finite for $0 < \alpha \le \beta$, it is finite for all $\alpha > 0$. Now

$$\alpha\hat{U}^\alpha 1(x) = \alpha\hat{E}^x \int_0^\zeta e^{-\alpha t}\, dt = \hat{E}^x(1 - e^{-\alpha\zeta})$$

decreases to $\hat{P}^x(\zeta = \infty)$ as α decreases to zero. Using (2.14) and (5.2) it is now a standard argument to show that

$$\alpha E_B(\alpha) \to \int \hat{P}^x(\zeta = \infty)\pi_B(dx) \equiv \gamma(B)$$

as α decreases to zero. It then follows from a standard Tauberian theorem [17] that $t^{-1}e_B(t) \to \gamma(B)$ as $t \to \infty$ establishing (2.16).

If $\hat{\eta}(x) \equiv \hat{P}^x(\zeta = \infty)$, then the Laplace-Stieltjes transform of $\int m(dx)\hat{\eta}(x)\hat{P}^x(T_B \le t)$ is given by $(\hat{\eta}, U^\alpha(\pi^\alpha_B)) = \int \hat{U}^\alpha\hat{\eta}\, d\pi^\alpha_B$. But it is immediate that $\alpha\hat{U}^\alpha\hat{\eta} = \hat{\eta}$ for $\alpha > 0$, and so (2.17) follows by (a simpler version of) the above argument. The corresponding results for S_B are proved in exactly the same manner.

(5.3) REMARKS. Let B satisfy the hypotheses of (2.16) and define $e^*_B(t) = \int m(dx)P^x(0 < T_B \le t)$. Then $e_B(t) = e^*_B(t) + m(B^r)$ where B^r is the set of regular points of B. Since $e_B(t)$ is finite, $m(B^r) < \infty$. But $B - B^r$ is semipolar and so $m(B - B^r) = 0$ [7, (6.10)]. Hence the fine closure of B and B itself must have finite m measure. More importantly, $t^{-1}e_B^*(t) \to \gamma(B)$ as $t \to \infty$. Using the fact that m is excessive it is readily checked that e_B^* is subadditive and so

$$\gamma(B) = \lim_{t\to\infty} t^{-1}e_B^*(t) = \inf_{t>0} t^{-1}e_B^*(t).$$

It is easy to see that

$$\lim_{t\to 0} t^{-1}e_B^*(t) = \sup_{t>0} t^{-1}e_B^*(t)$$

may be infinite. Consider, for example, Brownian motion on $]0,\infty[$ and take $B = \{1\}$. Of course, for this example, $\gamma(B) = 0$ because $\hat{P}^x(\zeta = \infty) = 0$.

6. Asymptotics II: Proof of (2.23), (2.26), and (2.28)

Note that the potential of K^B defined in (2.19) is given by

$$E^x[K_\infty^B] = P^x(L_B > 0) = \phi_B(x).$$

The proof of Theorem 2.23 is very similar to that of (A-14) in [5] to which we refer the reader. In particular the Revuz formula

$$U^\alpha(f\pi_B) = E^\cdot\left[e^{-\alpha L_B} f(X_{L_B-}); L_B > 0\right] \tag{6.1}$$

is valid for $f \geq 0$ and $\alpha \geq 0$.

The first assertion in (2.26) is just the definition of the Revuz measure and is contained already in (2.20). If $f \geq 0$ and $\pi_B(f) < \infty$, then $H_B(f,t)$ is finite for each t. Using (6.1)

$$\alpha \int_0^\infty e^{-\alpha t} d_t H_B(f,t) = \alpha \int m(dx) E^x[e^{-\alpha L_B} f(X_{L_B-}) : L_B > 0]$$

$$= \alpha(1, U^\alpha(f\pi_B)) = \int \alpha \hat{U}^\alpha 1 \, fd\pi_B$$

$$\to \int \hat{P}^x(\zeta = \infty) f(x) \pi_B(dx) \equiv \rho(f)$$

as $\alpha \to 0$, and consequently $t^{-1} H_B(f,t)$ approaches $\rho(F)$ as $t \to \infty$ establishing the second assertion in (2.26). For the last, note that because m is excessive $H_B(f,t)$ is subadditive in t — this is the key observation for developing Revuz measures, see [7], for example. As a result

$$\pi_B(f) = \sup_{t>0} t^{-1} H_B(f,t) \geq \inf_{t>0} t^{-1} H_B(f,t) = \rho(F).$$

But if $\hat{P}^x(\zeta = \infty) = 1$ a.e. π_B, $\rho(F) = \pi_B(f)$ completing the proof of (2.26).

To prove (2.28) first note that $\hat{\eta}(x) \equiv \hat{P}^x(\zeta = \infty)$ is coinvariant, that is, $\hat{P}_t \hat{\eta} = \hat{\eta}$ for all $t > 0$. It follows from this for $f \in \mathcal{E}^+$ that

$$G(t) \equiv \int m(dx) \hat{\eta}(x) E^x[f(X_{L_B-}); \ 0 < L_B \leq t]$$

is additive; $G(t + s) = G(t) + G(s)$ for $t,s \geq 0$. Since $G \geq 0$, $G(t) = Ct$ for some constant, $0 \leq C \leq \infty$. Now using (6.1) and $\hat{U}^\alpha \hat{\eta} = \alpha^{-1} \hat{\eta}$ for $\alpha > 0$,

$$\int_0^\infty e^{-\alpha t} G(t) dt = \frac{1}{\alpha} \int m(dx) \hat{\eta}(x) E^x\left[e^{-\alpha L_B} f(X_{L_B-}); \ 0 < L_B\right]$$

$$= \frac{1}{\alpha}(\hat{\eta}, U^\alpha(f\pi_B)) = \frac{1}{\alpha} \int \hat{U}^\alpha \hat{\eta} f d\pi_B = \frac{1}{\alpha^2} \int \hat{\eta} f \, d\pi_B,$$

and it follows that $C = \int \hat{\eta} f d\pi_B$ in all cases, proving (2.28).

7. Conditional Capacities

We begin by recalling some facts about h-transforms. We restrict ourselves to considering $(X,\hat{X},m)$ in weak duality as in the preceding sections. Let $h \in \mathcal{S}$; that is, h is a Borel measurable excessive function with $h < \infty$ a.e. m. Define $E_h = \{x: 0 < h(x) < \infty\}$ and

$$(7.1) \qquad P_t^h f(x) = h(x)^{-1} P_t(fh)(x) \qquad x \in E_h$$
$$= e^{-t} f(x) \qquad x \in E - E_h.$$

Then (P_t^h) is a semigroup on E with $P_t^h(x,\cdot) = e^{-t}\varepsilon_x$ if $x \in E - E_h$. This convention for $x \in E - E_h$ is somewhat unusual, but it is very convenient for weak duality. Note that it differs slightly from the convention adopted in [5], where $P_t^h(x,\cdot)$ was taken to be ε_x for all $t \geq 0$ if $x \in E - E_h$. The definition in (7.1) is more convenient here, since it follows that if X is transient so is X^h. Recall that Ω is the space of all right continuous functions from $\mathbb{R}^+$ to $E_\Delta = E \cup \{\Delta\}$ with Δ as cemetery and which have left limits in E on $]0,\zeta[$, and that $X_t(\omega) = \omega(t)$ are the coordinate maps. It is well known that there exist probabilities $P^{x/h}$ on $(\Omega,\mathcal{F}^0)$ such that $X^h = (X_t, P^{x/h})$ is a Borel right process on E with transition semigroup (P_t^h). Starting at $x \in E - E^h$, X^h remains at x for an exponential holding time with parameter one and then jumps to Δ. It is not difficult to check that $(X^h, \hat{X}, hm)$ is in weak duality. See [5, (5.4)]. We refer the reader to [5] or [16] for additional properties of h-transforms.

We can now state the characterization of potentials given in [5, (7.9)]. Recall the definition of Λ in (2.22). *A function* $u \in \mathcal{S}$ *is the potential of a measure* $\mu \in \mathfrak{M}$ *if and only if* $P^{x/u}(\Lambda) = 1$ a.e. m *on* E_u. Moreover it was shown in [5] that if $u,v \in \mathcal{S}$ with $u = v$ a.e. m, then $P^{\cdot/u} = P^{\cdot/v}$ a.e. m on $E_u \cap E_v$. (See the discussion below (6.5) in [5].) In particular, in (2.25) one may replace ϕ_B by $u \in \mathcal{S}$ with $u = \phi_B$ a.e. m without changing the condition. Hence, (2.25) characterizes transient sets.

The connection with capacities is based on the following well known property of h-transforms. See [16]. If T is an $(\mathcal{F}^*_{t+})$ stopping time, in particular, if $T = T_B$ or $T = S_B$, then

$$(7.2) \qquad P^{x/h}[T < \zeta] = [h(x)]^{-1} E^x[h(X_T)]$$

for $x \in E_h$. (Here $(\mathcal{F}^*_{t+})$ is the right continuous universal completion of the filtration $(\mathcal{F}^0_t)$.) We shall use the notation $U^{\alpha,h}$ for the resolvent of (P^h_t). Thus

$$\begin{aligned} U^{\alpha,h}f(x) &= h(x)^{-1}\, U^\alpha(fh)(x) \qquad x \in E_h \\ &= (\alpha+1)^{-1} f(x) \qquad x \in E - E_h. \end{aligned}$$

However, we write U^h for $U^{0,h}$. In writing potentials relative to the triple $(X^h, \hat{X}, hm)$ or more complicated triples it is better to use measures rather than functions in order to avoid confusion. For example, $U^h(\mu)$ is the Borel measurable h-excessive density of $\mu\hat{U}$ with respect to hm, provided it exists. Finally note that hm is carried by E_h since $m(\{h = \infty\}) = 0$.

(7.3) PROPOSITION. $B \in \mathcal{E}$ *is transient relative to* (X^h, X, hm) *if and only if* $P_B h$ *is the potential of a measure* $\mu \in \mathfrak{M}$ *(relative to* $(X, \hat{X}, m)$ *so that* $\mu\hat{U} = P_B h\cdot m$*), and then the capacitary measure* π^h_B *of* B *relative to* $(X^h, \hat{X}, hm)$ *equals* μ.

PROOF. Using (7.2) and the evident notation $\phi^h_B \equiv P^h_B 1 = h^{-1} P_B h$ on E_h. But h being excessive, $P_B h = 0$ on $\{h = 0\}$ so $P_B h = h\phi^h_B$ on

$\{h < \infty\}$. Since $m(\{h = \infty\}) = 0$ we see that $\mu\hat{U} = P_B h \cdot m$ if and only if $\mu\hat{U} = \phi_B^h \cdot hm$. □

(7.4) COROLLARY. *If* $h = U(\mu)$ *then every* $B \in \mathcal{E}$ *is transient and left transient relative to* $(X^h, \hat{X}, hm)$ *and* $\pi_B^h = \mu\hat{P}_{B-}$ *and* $\pi_{B-}^h = \mu\hat{P}_B$.

PROOF. Since $P_B h = P_B U(\mu) = U(\mu\hat{P}_{B-})$, the first assertion follows from (7.3). The second follows from the left version of (7.3). □

We now fix $h \in \mathcal{S}$ and $\hat{h} \in \hat{\mathcal{S}}$ and assume that both h and $\hat{h}$ are *finite*. Then $(X^h, \hat{X}^{\hat{h}}, h\hat{h}m)$ is in weak duality, and X^h and $\hat{X}^{\hat{h}}$ are transient. We write $\pi_B^{h,\hat{h}}$ for the capacitary measure of B relative to this triple whenever it exists. Thus $\pi_B = \pi_B^{1,1}$, π_B^h of (7.3) is $\pi_B^{h,1}$ and $\pi_B^{1,\hat{h}}$ is the capacitary measure of B with respect to $(X, \hat{X}^{\hat{h}}, \hat{h}m)$.

(7.5) PROPOSITION. *If* B *is transient (relative to* $(X, \hat{X}, m)$*), then it is transient relative to* $(X, \hat{X}^{\hat{h}}, \hat{h}m)$, *and* $\pi_B^{1,\hat{h}} = \hat{h}\,\pi_B$.

PROOF. If B is transient, $\phi_B m = \pi_B \hat{U}$ and so $\phi_B\,\hat{h}m = \pi_B\,\hat{U}\cdot\hat{h}$. Now $\hat{U}^\alpha\hat{h} \le \alpha^{-1}\hat{h}$ and so $\hat{U}^\alpha\hat{h}$ vanishes on $\{\hat{h} = 0\}$ if $\alpha > 0$. Letting $\alpha \to 0$ we see that $\hat{U}\hat{h} = 0$ on $\{\hat{h} = 0\}$. Therefore, because $\hat{h} < \infty$, if $f \in b\mathcal{E}^+$

$$(\hat{h}\pi_B)\hat{U}^{\hat{h}} f = \int \pi_B(dx)\hat{U}(f\hat{h})(x) = \int \phi_B \hat{h}\, f\, dm,$$

and so $(\hat{h}\pi_B)\hat{U}^{\hat{h}} = \phi_B\,\hat{h}\,m$. That is, $\pi_B^{1,\hat{h}} = \hat{h}\,\pi_B$. □

(7.6) COROLLARY. *If* B *is transient relative to* $(X^h, \hat{X}, hm)$, *then it is transient relative to* $(X^h, \hat{X}^{\hat{h}}, h\,\hat{h}\,m)$ *and* $\pi_B^{h,\hat{h}} = \hat{h}\,\pi_B^{h,1}$.

PROOF. Apply (7.5) to $(X^h, \hat{X}, hm)$.

Of course, we define $C^{h,\hat{h}}(B) = \pi_B^{h,\hat{h}}(1)$, $C_-^{h,\hat{h}}(B) = \pi_{B-}^{h,\hat{h}}(1)$, and so on whenever these quantities exist. The next proposition shows that $C^{h,\hat h}$ agrees with the capacities defined by Hunt in section 19 of [8]. It also gives an independent verification of $C^{h,\hat h}(B) = \hat C_-^{h,\hat h}(B)$ in the present situation.

(7.7) PROPOSITION. *Let* $h = U(\mu)$ *and* $\hat h = \hat U(\hat\mu)$ *be everywhere finite. Then*

$$C^{h,\hat h}(B) = \int \hat P_{B-}\,\hat h\, d\mu = \int P_B\, h\, d\hat\mu = \hat C_-^{h,\hat h}(B).$$

PROOF. From (7.4), (7.6) and their duals $\pi_B^{h,\hat h} = \hat h\cdot\mu\hat P_{B-}$ and $\hat\pi_{B-}^{h,\hat h} = h\cdot\hat\mu P_B$. Therefore

$$\begin{aligned} C^{h,\hat h}(B) &= \int \hat h\, d\mu\hat P_{B-} = \int \hat P_{B-}\hat h\; d\mu = \int \hat U(\hat\mu P_B) d\mu \\ &= \int U(\mu) d\hat\mu P_B = \int h\, d\hat\mu P_B = \int P_B\, h\, d\hat\mu. \end{aligned}$$

Since $\int h\, d\hat\mu P_B = \hat C_-^{h,\hat h}(B)$, (7.7) is established. □

(7.8) REMARK. Returning to (2.16) recall that $\gamma(B) = \int \hat P^x(\zeta = \infty)\pi_B(dx)$. But $\hat\eta(x) = \hat P^x(\zeta = \infty)$ is coexcessive (actually it is coinvariant, $\hat P_t\hat\eta = \hat\eta$), and so by (7.5) $\gamma(B) = C^{1,\hat\eta}(B)$ is the capacity of B relative to $(X, \hat X^{\hat\eta}, \hat\eta m)$. But $\hat X^{\hat\eta}$ is "just" $\hat X$ conditioned to have an infinite life-time.

(7.9) REMARK. Let q be a strictly positive Borel function with $m(q) < \infty$, $h \equiv Uq \le 1$, and $\hat h \equiv \hat U q \le 1$. Of course, $0 < h \le 1$ and $0 < \hat h \le 1$. From the point of view of Markov process theory $\Gamma(B) \equiv C^{h,\hat h}(B)$ is a much more reasonable definition of capacity than $C(B)$. For example, all sets $B \in \mathcal{E}$ are transient (and left transient) relative $(X^h, \hat X^{\hat h}, h\hat h m)$ so $\Gamma(B)$ exists for all B and is finite since q is integrable. From (7.4) and (7.5) the corresponding capacitary measures are given by $\pi_B^{h,\hat{h}} = \hat{h}(qm)\hat{P}_{B-}$ and $\pi_{B-}^{h,\hat{h}} = \hat{h}(qm)\hat{P}_B$. Note that this last measure is always carried by $B \cup B^{cr}$. Since h and $\hat{h}$ are strictly positive one still has $\Gamma(B) = 0$ (resp. $\Gamma_-(B) = 0$) if and only if B is m-polar (resp. left m-polar). Moreover $\Gamma(B)$ has many nice properties as a set function which we shall not discuss here. But see [8]. Perhaps this is what Hunt had in mind when he introduced these capacities in [8], albeit in quite a different manner.

8. The Stationary Process. Proof of (2.27)

In [7] following [9] and [11] we constructed a stationary process (Z,P) over $(X,\hat{X},m)$. Here Z_t is defined for $t \in \mathbb{R}$ and $Z_t \in E$ if and only if $\alpha(\omega) < t < \beta(\omega)$ where $\alpha \geq -\infty$ is the birth time of Z and $\beta \leq +\infty$ is the death time. On $]\alpha,\beta[$, $t \to Z_t$ is right continuous with left limits in E. The measure P on an appropriate path space is in general only σ-finite. The connection with $(X,\hat{X},m)$ is described as follows. The law of $(Z_{s+t})_{t\geq 0}$ under P on $\{\alpha < s < \beta\}$ is the same as that of $(X_t,P^m)_{t\geq 0}$ and the law of $(Z_{(s-t)-})_{t\geq 0}$ under P on $\{\alpha < s < \beta\}$ is the same as that of $(X_t,\hat{P}^m)_{t\geq 0}$. Loosely speaking this says that looking in the forward direction Z_{s+t} is a copy of X under P^m while looking backwards it is a copy of $X^- = (X_{t-})$ under $\hat{P}^m$ provided Z is alive at time s. In particular (Z,P) is stationary. We refer the reader to [7] for more details.

If $B \in \mathcal{E}$ define

$$\tau_B = \inf\{t: Z_t \in B\}, \qquad \sigma_B = \inf\{t: Z_{t-} \in B\},$$

$$\lambda_B = \sup\{t: Z_t \in B\}, \qquad \mu_B = \sup\{t: Z_{t-} \in B\},$$

where in these definitions the infimum of the empty set is $+\infty$ while the supremum of the empty set is $-\infty$. (Note the difference in the convention here.) In section 13 of [7] a number of formulas interpreting the various capacitary measures are given. We recall two of them. For (8.1), B is strongly left cotransient while for (8.2) it is strongly transient.

$$(8.1)\qquad P[\tau_B \in dt,\ Z(\tau_B) \in dx] = dt\ \hat{\pi}_{B-}(dx).$$

$$(8.2)\qquad P[\lambda_B \in dt,\ Z(\lambda_{B-}) \in dx] = dt\ \pi_B(dx).$$

Note that these give, using $C(B) = \hat{C}_-(B)$,

$$(8.3)\qquad P[\tau_B \in dt] = C(B)dt = P[\lambda_B \in dt].$$

We now turn to the proof of (2.27). Since B is strongly left cotransient it follows from (8.1) that

$$(8.4)\qquad P[f(Z_{\tau_B});\ 0 < \tau_B \le t] = t\hat{\pi}_{B-}(f).$$

Of course, f is taken to be zero off E so that $\{f(Z_{\tau_B}) > 0\} \subset \{\alpha < \tau_B < \beta\}$. The strong left cotransience of B means in terms of Z that $P[\tau_B \le \alpha]$ = 0. See [7]. Now the indicator of $]0,1[\ \cap\ ]\alpha,\infty[$ is the increasing limit as $n \to \infty$ of

$$\sum_{k=0}^{2^n-1} 1_{]k2^{-n},(k+1)2^{-n}]}\ 1_{\{\alpha < k2^{-n}\}},$$

and so from (8.4)

$$\hat{\pi}_{B-}(f) = \lim_n \sum_{k=0}^{2^n-1} P[f(Z_{\tau_B});\ k2^{-n} < \tau_B \le (k+1)2^{-n},\ \alpha < k2^{-n}].$$

But $f(Z_{\tau_B}) = 0$ unless $\tau_B < \beta$ and so one may replace $\alpha < k2^{-n}$ by $\alpha < k2^{-n} < \beta$ in this last expression. Since Z is stationary this becomes

$$\hat{\pi}_{B-}(f) = \lim_{n\to\infty} 2^n P[f(Z_{\tau_B}); \ 0 < \tau_B \le 2^{-n}, \ \alpha < 0 < \beta].$$

Next observe that, writing $a = 2^{-n}$ for convenience

$$\{0 < \tau_B \le a\} = \{0 < \tau_B \le a\} \cap \{Z_t \notin B \quad \text{for any } t \le 0\}.$$

Therefore, using the relationship between Z and $(X,\hat{X},m)$ — explicitly [7, (10.5)]

$$(8.5) \qquad \hat{\pi}_{B-}(f) = \lim_{n\to\infty} 2^n \int m(dx)\hat{P}^x(S_B = \infty)E^x[f(X_{T_B}); \ 0 < T_B \le 2^{-n}].$$

Let $u(t) = \int m(dx)\hat{P}^x(S_B = \infty)E^x[f(X_{T_B}); \ 0 < T_B \le t]$. Then (2.27) asserts that $u(t) = t\hat{\pi}_{B-}(f)$ while (8.5) states that $2^n u(2^{-n}) \to \hat{\pi}_{B-}(f)$ as $n \to \infty$. Thus to complete the proof of (2.27) it suffices to show that u is additive. To this end let (Q_t) (resp. $(\hat{Q}_t)$) be the semigroup of X (resp. $\hat{X}$) killed at T_B (resp. S_B). It can be shown that (Q_t) and $(\hat{Q}_t)$ are dual semigroups relative to m; that is, $(Q_t g,\hat{g}) = (g,\hat{Q}_t\hat{g})$ for $t \ge 0$ and $g,\hat{g} \in \mathcal{E}^+$. Let $\hat{h}(x) = \hat{P}^x(S_B = \infty)$. Then it is easily checked that $\hat{Q}_t\hat{h} = \hat{h}$. Let $h_t(x) = E^x[f(X_{T_B}); \ 0 < T_B \le t]$. Then $u(t) = (\hat{h},h_t)$. But

$$h_{t+s}(x) = h_t(x) + E^x[f(X_{T_B}); \ t < T_B \le t+s] = h_t(x) + Q_t h_s(x),$$

and so

$$u(t+s) = (\hat{h},h_t) + (\hat{h},Q_t h_s) = (\hat{h},h_t) + (\hat{Q}_t\hat{h},h_s) = u(t) + u(s),$$

completing the proof of (2.27).

We conclude this section with a sufficient condition for strong transience.

(8.6) PROPOSITION. *Suppose* $B \in \mathbf{E}$ *and* $B \subset \{Ug \geq 1\}$ *where* $g \in \mathbf{E}^+$ *and* $Ug < \infty$ a.e. m. *If there exists an increasing sequence* (ζ_n) *of stopping times such that* P^m *almost surely* $\zeta_n < \zeta$ *for each* n *and* $\lim \zeta_n = \zeta$, *then* B *is strongly transient.*

PROOF. Fix an x with $Ug(x) < \infty$ and (ζ_n) increasing to ζ strictly from below almost surely P^x. Then

$$P_{\zeta_n} Ug(x) \geq P_{\zeta_n + T_B \circ \theta_{\zeta_n}} Ug(x) \geq P^x(\zeta_n + T_B \circ \theta_{\zeta_n} < \zeta).$$

But $P_{\zeta_n} Ug(x) \to 0$ as $n \to \infty$ and consequently $\lim_{n\to\infty} P^x(\zeta_n + T_B \circ \theta_{\zeta_n} < \zeta) = 0$ for m almost all x. But if $L_B = \zeta$ and $\zeta_n < \zeta$, then $\zeta_n + T_B \circ \theta_{\zeta_n} < \zeta$. Consequently, $P^m(L_B = \zeta) = 0$. □

(8.7) REMARKS. If $\zeta = \infty$ a.s. P^m, then obviously $\zeta_n = n$ work in (8.6). Suppose X is a Hunt process on a locally compact state space E with a countable base and Δ is taken to be the point at infinity in the one point compactification of E. Suppose X has continuous trajectories on $[0,\zeta[$ and there is no killing in E, that is, $X_t \to \Delta$ as $t \uparrow \zeta$ when $\zeta < \infty$. If (G_n) is an increasing sequence of open sets with compact closures such that $\bar{G}_n \subset G_{n+1}$ and $\bigcup G_n = E$, then $\zeta_n = T_{G_n^c}$ work in (8.6). See also (2.10) in connection with (8.6).

9. Elementary Examples

(9.1) Let X and $\hat{X}$ be translation to the right and left respectively on $\mathbb{R}$ at unit speed. Then X and $\hat{X}$ are in strong duality with respect to Lebesgue measure m. The potential kernel u(x,y) in (1.5) is given by $u(x,y) = 1$ if $x < y$ and $u(x,y) = 0$ if $x \geq y$. Of course, X and $\hat{X}$ are even Hunt processes with continuous trajectories. Let $B =]-\infty,0]$. It is immediate that $\phi_B = 1_{]-\infty,0[}$, $\pi_B = \varepsilon_0$, and $\hat{P}^{\alpha}_{B-}(x,\cdot) = \varepsilon_x$ if $x < 0$. Thus from (4.4), $\pi^{\alpha}_B = \varepsilon_0 + \alpha\phi_B \cdot m$. Therefore π^{α}_B does not approach π_B as α approaches zero. In fact $C^{\alpha}(B) = \infty$ if $\alpha > 0$ and $C(B) = 1$. Moreover, $P^{\cdot}(T_B \le t) = 1_{]-\infty,0[}$ and so $\int m(dx)P^x(T_B \le t) = \infty$ for all $t \ge 0$. Consequently the condition $C^{\alpha}(B)$ finite for some $\alpha > 0$ in (2.16) cannot be replaced by $C(B) < \infty$.

(9.2) The state space consists of two copies of the $\mathbb{R}$, say $\mathbb{R}_0 = \mathbb{R} \times \{0\}$ and $\mathbb{R}_1 = \mathbb{R} \times \{1\}$. The process X starting on $\mathbb{R}_0$ at $(x,0)$ with $x < 0$ translates at unit speed to the right but jumps to $(0,1)$ on $\mathbb{R}_1$ as it approaches $(0,0)$. Starting at $(x,0)$ with $x \ge 0$ it just translates to right. The motion is defined analogously if it starts on $\mathbb{R}_1$ — it jumps to $(0,0)$ on $\mathbb{R}_0$ as it approaches $(0,1)$ from the left. $\hat{X}$ is a similar process but translating to the left and as it approaches zero on one line it jumps to zero on the other. If m is Lebesgue measure on each of the lines $\mathbb{R}_0$ and $\mathbb{R}_1$, then $(X,\hat{X},m)$ is in strong duality, but neither X nor $\hat{X}$ is standard. If $B = \{(0,1)\}$, then $\pi_B = \varepsilon_{(0,0)}$ is not carried by $\bar{B}$.

ADDED NOTE. P.J. Fitzsimmons has proved recently that the converse of (2.23) is valid. That is, if B is transient, then $\{L_B = \zeta\} \subset \Lambda$ a.s. P^m and π_B is given by (2.20), and the analagous statement for left transient sets is also valid. In addition, Fitzsimmons has shown that for any left transient set B, π_{B-} is carried by $B \cup B^{cr}$. See (2.12) and the ensuing discussion.

References

1. R.M. BLUMENTHAL and R.K. GETOOR. *Markov Processes and Potential Theory*. Academic Press, New York, 1968.

2. M. FUKUSHIMA. Dirichlet spaces and strong Markov processes. *Trans. Amer. Math. Soc. 162* (1971), 185-224.

3. R.K. GETOOR. Some asymptotic formulas involving capacity. *Zeit. f. Wahrs. verw. Geb. 4* (1965), 248-252.

4. R.K. GETOOR. Transience and recurrence of Markov processes. Sém. de Prob. XIV. Lecture Notes in Math. *784*. Springer, 1980.

5. R.K. GETOOR and J. GLOVER. Riesz decompositions in Markov process theory. To appear in *Trans. Amer. Math. Soc.*

6. R.K. GETOOR and M.J. SHARPE. Last exit times and additive functionals. *Ann. Prob. 1* (1973), 550-569.

7. R.K. GETOOR and M.J. SHARPE. Naturality, standardness, and weak duality for Markov processes. Submitted to *Zeit. f. Wahrs. verw. Geb.*

8. G.A. HUNT. Markov processes and potentials III. *Ill. J. Math. 2* (1958), 151-213.

9. S.E. KUZNETSOV. Construction of regular split processes. *Theory Prob. and Appl. 22* (1977), 773-793.

10. P.A. MEYER, R.T. SMYTHE, and J.B. WALSH. Birth and death of Markov processes. *Proc. 6th Berk. Symp. Stat. Prob.*, Vol. III, 295-305. Univ. of Cal. Press, 1972.

11. J.B. MITRO. Dual Markov processes: construction of a useful auxiliary process. *Zeit. f. Wahrs. verw. Geb. 47* (1979), 97-114.

12. S.C. PORT and C.J. STONE. Infinitely divisible processes and their potential theory. Part I. *Ann. Instit. Fourier, Grenoble 21* (1971), 157-275.

13. M.L. SILVERSTEIN. *Symmetric Markov Processes*. Lecture Notes in Math. *426* (1974). Springer.

14. F. SPITZER. Electrostatic capacity, heat flow, and Brownian motion. *Zeit. f. Wahrs. verw. Geb. 3* (1964), 110-121.

15. J.B. WALSH. Markov processes and their functionals in duality. *Zeit. f. Wahrs. verw. Geb. 24* (1972), 229-246.

16. J.B. WALSH. The cofine topology revisited. *Proc. Symp. in Pure Math.* *31* (1977), 131-152. Ed., J.L. Doob. AMS Providence.

17. D.V. WIDDER. *The Laplace Transform.* Princeton Univ. Press, 1946.

R.K. GETOOR
Department of Mathematics
University of California-San Diego
La Jolla, CA 92093

Seminar on Stochastic Processes, 1983
Birkhäuser, Boston, 1984

RAY-KNIGHT'S THEOREM ON BROWNIAN LOCAL TIMES AND TANAKA'S FORMULA

by

T. JEULIN

Using Tanaka's formula in an appropriate filtration, we give a representation property for the excursions below a given level of the (possibly killed) Brownian motion. Ray-Knight's theorems on Brownian local times are then directly deduced from Tanaka's formula.

1. Introduction

Let X be a real valued Brownian motion defined on some complete probability space $(\Omega,\mathbf{A},\mathbb{P})$. The filtration generated by X will be denoted by $(\mathcal{X}_t)_{t\geq 0}$.

Local time

For real x, its positive and negative parts are denoted by x^+ and x^-. Recall *Tanaka's formula*, which may be interpreted as a definition of the local times L_t^a:

$$(X_t-a)^- = (X_0-a)^- - \int_0^t 1_{\{X_s\leq a\}}\, dX_s + \tfrac{1}{2}L_t^a.$$

For fixed a, $t \to L_t^a$ is increasing and is supported by $\{t:X_t = a\}$.

Further, Trotter's theorem ensures the existence of a continuous version of $(a,t) \to L_t^a$.

Local time is an *occupation time density*: for $t \geq 0$ and f Borel measurable,

$$\int_0^t f(X_s)\, ds = \int f(a)\, L_t^a\, da.$$

Finally, we have Ito-Tanaka's general formula: let F be a convex function on $\mathbb{R}$, F'_g its left derivative and $F''(dx)$ its second derivative; then

$$F(X_t) = F(X_0) + \int_0^t F'_g(X_s)\, dX_s + \tfrac{1}{2} \int L_t^a\, F''(da).$$

Excursions

Let $x \in \mathbb{R}$ and K a positive random variable. Define

$$C_t(x,K) = \int_0^{t\wedge K} 1_{\{X_s \leq x\}}\, ds,$$

the time spent below the level x up to time $\inf(t,K)$, and let $\tau_t(x,K)$ be its right continuous inverse. Following D. Williams [13], let $\mathcal{E}_x(K)$ be the σ-field of the *excursions of* X *below the level* x *on the time interval* $[0,K]$: with $X_\infty = \partial$, where ∂ is some cemetery,

$$\mathcal{E}_x(K) = \sigma(X_{\tau_t(x,K)},\ t \geq 0).$$

It is easily seen that $x \to \mathcal{E}_x(K)$ is increasing.

2. Recurrent Case

In this paragraph, we take $K = +\infty$, and write $\mathcal{E}_x$ instead of $\mathcal{E}_x(\infty)$. Let $F \in L^2(\mathfrak{X}_\infty)$. The *predictable representation property* for Brownian motion asserts that

$$F = \mathbb{E}[\, F \mid X_0 \,] + \int_0^\infty f_s \, dX_s,$$

for some predictable process f (with respect to the filtration $\mathfrak{X}$) such that

$$\mathbb{E}[\int_0^\infty f_s^2 \, ds] < \infty.$$

Moreover, if F is $\mathcal{E}_x$-measurable, we can be more precise:

LEMMA 1. *Let* $F \in L^2(\mathcal{E}_x)$. *There exists an* $\mathfrak{X}$*-predictable process* f *such that*

$$\mathbb{E}[\int_0^\infty f_s^2 \, ds] < \infty$$

$$F = \mathbb{E}[F \mid \inf(x,X_0)] + \int_0^\infty f_s \, 1_{\{X_s \le x\}} \, dX_s.$$

PROOF (adapted from McGill [6]). Write τ_t instead of $\tau_t(x,\infty)$ and let

$$N_t = (X_0 - x)^- - \int_0^{\tau_t} 1_{\{X_s \le x\}} \, dX_s.$$

Then, N is a continuous $(\mathfrak{X}_{\tau_t})_{t \ge 0}$ - martingale whose increasing process is

$$\int_0^{\tau_t} 1_{\{X_s \leq x\}} \, ds = t.$$

Thus, N is in fact an $(\mathfrak{X}_{\tau_t})_{t\geq 0}$ - Brownian motion. By Tanaka's formula applied at time τ_t, $t \to X_{\tau_t}$ satisfies Skorokhod's reflection equation:

$$x - X_{\tau_t} = N_t + \tfrac{1}{2} L^x_{\tau_t},$$

from which it is easily deduced that $L^x_{\tau_t} = 2 \sup_{s\leq t} N^-_s$ (see [2]) and, therefore, the filtrations generated by $(X_{\tau_t}, t\geq 0)$ and $(N_t, t\geq 0)$ are identical (up to null-sets). Making use of the predictable representation property, for $F \in L^2(\mathfrak{E}_x)$ there exists a process $\bar{f}$, predictable with respect to the filtration of N, such that

$$\mathbb{E}[\int_0^\infty \bar{f}^2_s \, ds] < +\infty \quad \textit{and} \quad F = \mathbb{E}[F \mid N_0] + \int_0^\infty \bar{f}_s \, dN_s.$$

But, using time change,

$$\int_0^\infty \bar{f}_s \, dN_s - \int_0^\infty \bar{f}_{C_s} 1_{\{X_s \leq x\}} \, dX_s,$$

and $f_. = \bar{f}_{C_.}$ is $\mathfrak{X}$-predictable. □

LEMMA 2. *Let* $x \in \mathbb{R}$ *and* $\mathfrak{G}^x_t = \bigcap_{\varepsilon>0} \sigma(\mathfrak{X}_{t+\varepsilon}, \mathfrak{E}_x)$. *For any* $\mathfrak{X}$-*local-martingale* Y, *the process*

$$(\int_0^t 1_{\{X_s > x\}} \, dY_s, \; t\geq 0)$$

is still a local-martingale with respect to the filtration $(\mathfrak{G}^x_t)_{t\geq 0}$.

PROOF. We may suppose $Y_0 = 0$ and $\sup_t |Y_t| \in L^1$. Let

$$Z_{\cdot} = \int_0^{\cdot} 1_{\{X_s \le x\}} \, dY_s, \quad 0 < s < t.$$

Let F be $\mathcal{E}_x$-measurable, g be $\mathcal{X}_s$-measurable (F and g bounded). Then, with f given by Lemma 1,

$$\mathbb{E}[(Z_t - Z_s)\, g\, F] = \mathbb{E}[g \int_s^t 1_{\{X_u > x\}} \, dY_u \cdot \int_0^{\infty} f_u \, 1_{\{X_u \le x\}} \, dX_u]$$

$$= \mathbb{E}[g \int_s^t f_u \, 1_{\{X_u > x, X_u \le x\}} \, d\langle Y,X\rangle_u] = 0.$$

Therefore, Z is a $(\mathcal{G}_t^x)_{t \ge 0}$ - martingale.

COROLLARY 3. *Let* $a \in \mathbb{R}$ *and let* T *be a* $(\mathcal{G}_t^a)_{t\ge 0}$ *- stopping time such that* $X_T \le a$ *a.s. and* L_T^a *is* $\mathcal{E}_a$*-measurable. Then,*

$$(L_T^y + 2(X_0 - y)^+, \quad y \ge a)$$

is a $(\mathcal{E}_y)_{y \ge a}$ *- continuous martingale whose increasing process is* $4 \int_a^y L_T^z \, dz$.

PROOF. i) It is easy to see that for all $y \ge a$, L_T^y is $\mathcal{E}_y$-measurable. It suffices now to prove that $(\tfrac{1}{2} L_T^y + (X_0 - y)^+, \ y \ge a)$ is a $(\mathcal{E}_y)_{y \ge a}$ - martingale. Its increasing process must then be $\int_a^{\cdot} L_T^z \, dz$, indeed, Bouleau and Yor [1] proved, as a consequence of Tanaka's formula, that for any positive finite random variable K, the quadratic variation of $x \to \tfrac{1}{2} L_K^x$ over the (space) interval $[u,v]$ is $\int_u^v L_K^z \, dz$.

ii) Let $a \le x < y$ and

$$V_t = \tfrac{1}{2}(L_t^y - L_t^x) + (X_0-y)^+ - (X_0-x)^+ - (X_t-y)^+ + (X_t-x)^+.$$

It remains to show that

$$\mathbb{E}[\ |V_T|\ |\ \mathcal{E}_x] < +\infty \quad \textit{and} \quad \mathbb{E}[\ V_T\ |\ \mathcal{E}_x] = 0.$$

But, from Tanaka's formula,

$$V_t = \int_0^t 1_{\{x<X_s\le y\}}\, dX_s,$$

and by Lemma 2, $(V_t)_{t\ge 0}$ is a $(\mathcal{G}_t^x)_{t\ge 0}$ - local-martingale. Thus, if $\mathbb{E}[\langle V,V\rangle_T\ |\ \mathcal{E}_x]$ is finite, conditionally on $\mathcal{E}_x$ $(\subset \mathcal{G}_0^x)$, $(V_{t\wedge T},\ t\ge 0)$ is in fact a $(\mathcal{G}_t^x)_{t\ge 0}$ - martingale (bounded in L^2) and $\mathbb{E}[V_T\ |\ \mathcal{G}_0^x] = V_0 = 0$.

In order to prove the latter condition, note that

$$\langle V,V\rangle_t = \int_0^t 1_{\{x<X_s\le y\}}\, ds.$$

Consider, for $p > 0$, $F(z) = \cosh\sqrt{2p}\ ((y-\sup(z,x))^+)$, $2c = \sqrt{2p}\tanh\sqrt{2p}\ (y-x)$, and let

$$W_t = F(X_{t\wedge T})\exp[-c(L_T^x - L_{t\wedge T}^x)]\ \exp[-p\int_0^{t\wedge T} 1_{\{x<X_s\le y\}}\, ds].$$

From Ito's formula, W is the bounded $(\mathcal{G}_t^x)$ - martingale

$$W_0 + \int_0^{t\wedge T} W_s\, 1_{\{x<X_s\le y\}}\, \frac{F_g'}{F}(X_s)\, dX_s.$$

Thus $\mathbb{E}[W_T\ |\ \mathcal{G}_0^x] = W_0$ and

$$\mathbb{E}[\exp - p\langle V,V\rangle_T\ |\ \mathcal{G}_0^x] = \frac{\cosh\sqrt{2p}(y-\sup(x,X_0))^+}{\cosh\sqrt{2p}\ (y-x)}\ \exp(-\frac{L_T^x}{2}\sqrt{2p})\tanh\sqrt{2p}\ (y-x).$$

It follows, by derivation at $p = 0$, that

$$\mathbb{E}[\langle V,V\rangle_T \mid \mathcal{G}_0^x] = (y-x)L_T^x + (y-x)^2 - ((y-\sup(x,X_0))^+)^2. \qquad \square$$

Now suppose $X_0 = 0$ and let T be as in Corollary 3 (for example $T = \inf(t{:}X_t=a)$ or $\inf(t{:}L_t^0 \geq \ell)$ for some $\ell > 0$ (take then $a = 0$), or T any "$\mathcal{E}_a$-representable" variable as defined by Walsh [11]). Then, $(L_T^{y+a}, y\geq 0)$ is an (inhomogeneous) strong Markov process (with respect to the filtration $(\mathcal{E}_{y+a}, y\geq 0)$ with generator $2\ell \frac{d^2}{d\ell^2} + 2\, 1_{\{y<a^-\}} \frac{d}{d\ell}$. It is also a Bessel process of dimension 2 for $0 \leq y \leq a^-$ and of dimension 0 for $y \geq a^-$. By time change, the same result is still true if we replace the Brownian motion X by any regular diffusion on a closed interval of $\mathbb{R}$ (in natural scale). This is Ray-Knight-Walsh's theorem in the recurrent case (see [5], [10] and [11]).

3. Transient Case

Let Y be a *transient regular diffusion* on an interval J of $\mathbb{R}$, with lifetime ζ and local time $(L_t^a, a\in J, t\in\mathbb{R}_+)$: assume Y is taken in *natural scale*. Via time change, the study of $(L_\zeta^a, a\in J)$ amounts to that of $(L_K^a, a\in J)$, where K is a killing time (of the Brownian motion X) associated with some *killing measure* μ.

Although the following results remain true in the general situation, we assume here, in order to avoid discussion about boundaries, that μ is a Radon measure on $\mathbb{R}$. Let U be an exponential random variable with rate 1, independent of X, and take from now on

$$K = \inf(t{:} \int L_t^a\, \mu(da) \geq U).$$

We first look for the analogue of Lemma 1 or 2 in the present situation. It appears that the "right" σ-fields to deal with are not the $\mathbf{E}_x(K)$'s but

$$\tilde{\mathbf{E}}_x(K) = \sigma(\mathbf{E}_x(K),\ X_0,\ X_K,\ I_K), \quad \textit{where} \quad I_K = \inf_{0\le s\le K} X_s.$$

In some sense, the first step includes Millar-Williams' decomposition of a transient diffusion at its minimum (see [8], [12]) and Pitman's theorem about the Bessel process of dimension 3 in [9]. Our method of proof uses enlargement of filtrations (see [4]).

PROPOSITION 4. *Let* V *be a solution of the equation*

$$V''(dx) - 2V(x)\,\mu(dx) = 0, \quad V > 0$$

on $\mathbb{R}$ *non-increasing. Let* $J_t = \inf_{s\ge t} X_{s\wedge K}$ *and let* $\mathbb{H}$ *be the smallest right-continuous filtration such that* $\mathbb{H}_t \supset \mathbb{X}_t$ *for all* t, K *is a* $\mathbb{H}$*-stopping time,* X_K *is* $\mathbb{H}_0$*-measurable,* J *is* $\mathbb{H}$*-adapted (note that* $J_0 = I_K$*). Then*

i) $\eta = \inf\{t: X_t = I_K\}$ *is the only time at which* X *attains its minimum on* $[0,K]$; *moreover,* $K = \inf(t\ge\eta: J_t = X_K)$.

ii) $\bar{X}_t = X_t - 2(J_t - J_0) - \int_0^{t\wedge K} \frac{V'}{V}(X_s)\,ds$ *defines a* $\mathbb{H}$*-Brownian motion.*

LEMMA 1'. *Let* $x \in \mathbb{R}$ *and* $F \in L^2(\tilde{\mathbf{E}}_x(K))$. *There exists a* $\mathbb{H}$*-predictable process* f *such that*

$$\mathbb{E}[\int_0^\infty f_s^2\,ds] < +\infty \quad \textit{and} \quad F = \mathbb{E}[F \mid X_0, X_K, I_K] + \int_0^K f_s\, 1_{\{X_s\le x\}}\, d\bar{X}_s.$$

SKETCH OF THE PROOF. Let

$$\tau_t = \tau_t(x,\infty), \quad \alpha = \int_0^K 1_{\{X_s \le x\}}\, ds,$$

$$\check{X}_t = X_{\tau_{t\wedge\alpha}}, \quad \check{J}_t = J_{\tau_{t\wedge\alpha}}, \quad \check{L}_t = L^x_{\tau_{t\wedge\alpha}}.$$

Then $\check{J}_t = \inf_{s\ge t} \check{X}_s$, $\alpha = \inf(t: \check{J}_t \ge \inf(x,X_K))$ and $\tilde{\mathbf{E}}_x(K)$ is the σ-field generated by X_0, X_K, I_K and the process $t \to \check{X}_t$. Moreover, from Tanaka's formula (in the filtration $\mathcal{H}$, at time $\tau_{t\wedge\alpha}$),

$$\check{X}_t = \inf(x,X_0) + 2(\check{J}_t - \check{J}_0) + \check{N}_t + \int_0^{t\wedge\alpha} \frac{V'}{V}(\check{X}_s)\, ds - \tfrac{1}{2}\check{L}_t,$$

where

$$\check{N}_t = \int_0^{\tau_{t\wedge\alpha}} 1_{\{X_s \le x\}}\, d\bar{X}_s$$

is a $(\mathcal{H}_{\tau_t})_{t\ge 0}$ - Brownian motion stopped at time α; $\check{X}_t \le x$; $\check{L}_t$ is non decreasing and supported by $\{t: \check{X}_t = x\}$.

Given $(\check{N}_\cdot, X_0, X_K, I_K)$, the previous equation has a unique solution (see [4]). $\tilde{\mathbf{E}}_x(K)$ is thus contained in $\sigma((\check{N}_t, t\ge 0), X_0, X_K, I_K)$. To conclude, we apply the predictable representation property for $\check{N}$ and use a time change (as in the proof of Lemma 1). □

With the same proof, Lemma 2 now becomes:

LEMMA 2'. *Let* $x \in \mathbb{R}$ *and* $\mathcal{H}^x_t = \bigcap_{\varepsilon>0} \sigma(\mathcal{H}_{t+\varepsilon}, \tilde{\mathbf{E}}_x(K))$. *Then, for any* $\mathcal{H}$*-local martingale* Z,

$$\left(\int_0^t 1_{\{X_s > x\}}\, dZ_s,\ t\ge 0\right)$$

is still a $(\mathbb{H}_t^X)_{t\geq 0}$*-local martingale.* □

Now let $x \leq y$ and apply Tanaka's formula in the filtration $(\mathbb{H}_t^X)_{t\geq 0}$ at time K to obtain:

$$(X_K-y)^+ - (X_K-x)^+ = (X_0-y)^+ - (X_0-x)^+ + \int_0^K 1_{\{x<X_s\leq y\}}\, d\bar{X}_s - 2\int_0^K 1_{\{x<X_s\leq y\}}\, dJ_s$$

$$- \int_0^K 1_{\{x<X_s\leq y\}} \frac{V'}{V}(X_s)\, ds + \tfrac{1}{2}(L_K^y - L_K^x).$$

By the occupation time formula,

$$\int_0^K 1_{\{x<X_s\leq y\}} \frac{V'}{V}(X_s)\, ds = \int_x^y \frac{V'}{V}(z)\, L_K^z\, dz.$$

Since J is non decreasing and is supported by $\{t:\ X_t = J_t\}$,

$$2\int_0^K 1_{\{x<X_s\leq y\}}\, dJ_s = 2\int_0^K 1_{\{x<J_s\leq y\}}\, dJ_s = 2\int_{I_K}^{X_K} 1_{\{x<z\leq y\}}\, dz$$

$(X_K = J_K,\ J_0 = I_K)$.

At last, let $p > 0$ and $G(z) = \dfrac{\cosh\sqrt{2p}((y-z)^+)}{V(z)}$. By Ito's formula,

$$\left(\exp\left[-p\int_0^{t\wedge K} 1_{\{X_s\leq y\}}\, ds\right] \frac{G(X_{t\wedge K})}{G^2(J_t)},\ t \geq 0\right)$$

is a $(\mathbb{H}_t)_{t\geq 0}$ - local martingale and is bounded conditionally on (X_0, X_K, I_K). Therefore,

$$\mathbb{E}\left[\exp\left[-p\int_0^K 1_{\{X_s\leq y\}}\, ds\right] \,\Big|\, X_0, X_K, I_K\right] = \frac{G(X_0)G(X_K)}{G^2(I_K)},$$

and by derivation at $p = 0$,

$$\mathbb{E}[\int_0^K 1_{\{X_s\le y\}} ds \,|\, X_0, X_K, I_K] = \frac{V^2(I_K)}{V(X_0)V(X_K)} (2[(y-I_K)^+]^2 - [(y-X_0)^+]^2 - [(y-X_K)^+]^2).$$

From Lemma 2',

$$(\int_0^{t\wedge K} 1_{\{x<X_s\le y\}} \, d\bar{X}_s, \ t \ge 0)$$

is then a L^2-bounded $(\mathbf{H}_t^x)_{t\ge0}$-martingale and

$$\mathbb{E}[\int_0^K 1_{\{x<X_s\le y\}} \, d\bar{X}_s \,|\tilde{\mathbf{E}}_x(K)] = 0 .$$

This proves the following:

COROLLARY 3'. *The process*

$$\frac{1}{2} L_K^x - 1_{\{x\ge I_K\}} \int_{I_K}^x (1_{\{y\le X_0\}} + 1_{\{y\le X_K\}} + \frac{V'}{V}(y) \, L_K^y) \, dy = \int_0^K 1_{\{X_s\le x\}} \, d\bar{X}_s$$

is a $(\tilde{\mathbf{E}}_x(K))_{x\in\mathbb{R}}$*-martingale, with increasing process* $(\int_{-\infty}^x L_K^y \, dy, \ x\in\mathbb{R})$. □

In particular, conditionally on (X_0, X_K, I_K), $(L_K^{y+I_K}, y\ge0)$ is an inhomogeneous strong Markov process (with respect to the filtration $(\mathbf{E}_{y+I_K}(K), \ y\ge0)$ with generator

$$2\ell \frac{d^2}{d\ell^2} + 2(1_{\{y\le X_0-I_K\}} + 1_{\{y\le X_K-I_K\}} + \frac{V'}{V}(y)\ell) \frac{d}{d\ell}$$

(see [5], [10] or [11]).

REMARK. Applications of the above technique to Perkin's theorem and related results are developed in [4].

References

1. N. BOULEAU et M. YOR. Sur la variation quadratique des temps locaux de certaines semi-martingales. *CRAS Paris, t. 292, Série I* (1981), 491-494.

2. M. CHALEYAT-MAUREL and N. EL KAROUI. Un problème de réflexion et ses applications au temps local et aux équations différentielles stochastiques sur $\mathbb{R}$; cas continu. *Astérisque 52-53* (1978), 117-144.

3. K. ITO and H.P. McKEAN. *Diffusion processes and their sample paths.* Springer-Verlag, Berlin, 1964.

4. T. JEULIN. Application de la théorie du grossissement à l'étude des temps locaux browniens. To appear.

5. F.B. KNIGHT. Random walks and a sojourn density of Brownian motion. *Trans. Amer. Math. Soc. 107* (1963), 56-86.

6. P. McGILL. Markov properties of diffusion local times; a martingale approach. *Adv. Appl. Prob. 14* (1982), 789-810.

7. P.A. MEYER. Un cours sur les intégrales stochastiques. *Séminaire de Probabilités X, Lect. Notes in Math. 511*, 245-400. Springer, Berlin, 1976.

8. P.W. MILLAR. A path decomposition for Markov processes. *Ann. Prob. 6* (1978), 345-348.

9. J.W. PITMAN. One dimensional Brownian motion and the three-dimensional Bessel process. *Adv. Appl. Prob. 7* (1975), 511-526.

10. D.B. RAY. Sojourn times of diffusion processes. *Ill. J. Math 7* (1963), 615-630.

11. J.B. WALSH. Excursions and local time. *Astérisque 52-53* (1978), 159-192.

12. D. WILLIAMS. Path decomposition and continuity of local time for one dimensional diffusions. *Proc. London Math. Soc. 28* (1974), 738-768.

13. D. WILLIAMS. Conditional excursion theory. *Séminaire de Probabilités XII, Lect. Notes in Math. 721*, 490-494. Springer, Berlin, 1979.

T. JEULIN
Universite Paris 7
2, Place Jussieu, Tour 45-55, 5e etage
75251 Paris Cedex 05, FRANCE

Seminar on Stochastic Processes, 1983
Birkhäuser, Boston, 1984

FURTHER RESULTS ON ENERGY

by

Z.R. POP-STOJANOVIC and MURALI RAO

1. Introduction

This is a continuation of earlier papers [5], [6] which treat the role of *energy* in probabilistic potential theory. Unfortunately, this role has not yet received the attention it enjoyed in classical potential theory where the *symmetry* of the potential kernel plays an essential role.

Here, as in earlier papers dealing with the same subject, we deal with the non-symmetric case. However, instead of the definition in our earlier papers, we shall use P.A. Meyer's definition of energy given for the supermartingale case [4] in order to define the energy of a class (D) potential. Our present approach gives an unified picture of this matter.

2. Energy of Class (D) Potentials

In order to use Meyer's definition we need to define additive functionals of class (D) potentials which may assume infinite values. This is the first thing we do.

Let $X = (\Omega, \mathcal{F}, \mathcal{F}_t, X_t, \theta_t, P^x)$ denote a transient Hunt process on a locally compact second countable state space with transition semigroup (P_t). We shall say that an almost everywhere finite excessive function s belongs to class (D) if, whenever T_n is a sequence of stopping times increasing to infinity, $P_{T_n} s$ decreases to zero almost everywhere. We emphasize that we do not require s to be a finite excessive function.

We will need to use additive functionals of class (D) potentials. Since we have allowed infinities for s we define this object as follows: Write

$$s = \sum s_n$$

where each s_n is bounded. It is well known that this can be done. Let A^n be the additive functional of s_n. It is then natural to put

$$A_t = \sum_n A_t^n \tag{1}$$

and call A the additive functional of s. Let us justify this definition. If A_∞ is finite then A_t is right continuous, $A_0 = 0$ and

$$A_{t+s} = A_t + A_s(\theta_t) \quad . \tag{2}$$

Also $E^x[A_\infty] < \infty$ if $s(x) < \infty$. If the latter happens we have

$$A_t + s(X_t) = E^x[A_\infty | \mathcal{F}_t]. \tag{3}$$

Further, if $s(x) < \infty$, P^x-almost surely, A and X can not have common discontinuities and hence, because of (3), A is uniquely determined. Here, we use the uniqueness of Doob-Meyer decomposition.

Now we can define the energy of a class (D) potential s. Let A be its additive functional as defined above. This is defined off the set where $s = \infty$, i.e., off a polar set. Write

$$(4) \qquad p(x) = E^x[A_\infty^2] .$$

We shall call p *the energy function of* s. Using (2) we can verify that p is a class (D) potential provided that it be finite almost everywhere.

For a class (D) potential q denote by $L(q)$ the total mass of the Revuz measure of q.

DEFINITION. *We say that a class* (D) *potential* s *is of* finite energy *if its energy function* p *is finite almost everywhere and has finite Revuz measure. We put*

$$\|s\|_e^2 = L(p).$$

More generally, if s *and* t *are class* (D) *potentials of finite energy with additive functionals* A *and* B, *we define the* mutual energy *of* s *and* t *by the formula*

$$\langle s,t \rangle = L(q)$$

where q *is the class* (D) *potential* $E^{\cdot}[A_\infty B_\infty]$.

It is a matter of verification that if s is the potential of a function f, that is, if $s = Uf$, then

$$\|s\|_e^2 = 2(s,f)$$

so that our definition agrees with the usual one.

The inner product $\langle \cdot,\cdot \rangle$ can now be extended on the linear space $\mathcal{R}$ of differences of potentials of finite energy. For $r \in \mathcal{R}$, $\|r\|_e = 0$

does not necessarily imply $r = 0$. However very simple conditions can be given to guarantee that $\|\cdot\|_e$ is indeed a norm. We will assume in the sequel that this is the case.

PROPOSITION 1. *Let* s *and* t *be excessive and have finite energy. Let* μ,ν *denote the Revuz measures of* s *and* t. *Then,*

$$(5) \qquad (\nu,s) + (\mu,t) \le 2\|s\|_e \, \|t\|_e$$

where (ν,s) *and* (μ,t) *denote integrals of* s *and* t *relative to* ν *and* μ *respectively.*

PROOF. Let A and B be the additive functionals of s and t respectively. We have

$$U_B s + U_A t = 2\, E^{\cdot}[A_\infty B_\infty] - E^{\cdot}[\int_0^\infty A_t \, dB_t] - E^{\cdot}[\int_0^\infty B_t dA_t] \le E^{\cdot}[A_\infty B_\infty],$$

where "integration by parts," Meyer [4], p. 114, is used. Since the Revuz measures of potentials $U_B s$ and $U_A t$ are $s\,d\nu$ and $t\,d\mu$ respectively, the proof of (5) is clear. □

REMARK. The proof of the above proposition shows that, if at least one of s and t is regular, then

$$(6) \qquad \langle s,t\rangle = (\nu,s) + (\mu,t) .$$

PROPOSITION 2. *A family* $\{s_\alpha\}$ *of excessive functions that is bounded in energy is uniformly integrable relative to the Revuz measure of every class* (D) *potential of finite energy.*

PROOF. Suppose $\|s_\alpha\|_e \le 1$. Let s be an excessive function of finite energy whose Revuz measure is μ and whose additive functional is A. For any measurable set B write $s_B = U_A(1_B)$. It is easy to see that as $\mu(B)$ tends to zero, so does $\|s_B\|_e$. Then (5) tells us that the integrals (μ, s_α) are uniformly bounded and that if $\mu(B)$ is small so is

$$(\mu, 1_B s_\alpha)$$

uniformly in α. □

If B is a Borel set such that $P_B 1$ is in class (D), the total mass of the Revuz measure of $P_B 1$ will be called the *capacity of B* and denoted by $C(B)$. With this definition we have the following.

PROPOSITION 3. *Let s be an excessive function and assume that s is of finite energy. Let $\lambda > 0$ and $B = (s > \lambda)$. Then,*

$$\lambda^2 \, C(B) \leq 4 \, \|s\|_e^2 \, . \tag{7}$$

PROOF. We can find a sequence of functions g_n vanishing off B and such that $s_n = Ug_n$ increases to $P_B 1$. Clearly, $\lambda P_B 1 \leq s$. Then it follows that $\lambda \|s_n\|_e \leq 2\|s\|_e$. Using (5) and the fact that the g_n vanish off B one has

$$\lambda \, (g_n, 1) + (s_n, \mu) \leq \frac{4}{\lambda} \, \|s\|_e^2 \, . \tag{8}$$

As $n \to \infty$, $(g_n \; 1)$ increases to $C(B)$. We obtain from (8)

$$\lambda \, C(B) + (P_B 1, \mu) \leq \frac{4}{\lambda} \, \|s\|_e^2 \, .$$

and (7) follows. □

The following energy formula of Meyer will be very useful below.

Let $s \in \mathfrak{R}$ and let A be the additive functional of s. Then

(9) $$E^x[A_\infty^2] = E^x[\int_0^\infty (s(X_t) + s(X_t)_-)\, dA_t].$$

More generally if $r, s \in \mathfrak{R}$ with additive functionals A, B respectively,

$$E^\cdot[A_\infty B_\infty] = E^\cdot[\int_0^\infty (r(X_t)_-\, dB_t + s(X_t)\, dA_t)].$$

Using (9) and following Meyer-Dellacherie [2], p. 175, we see that if $r, s \in \mathfrak{R}$, $r \le s$, then

(10) $$E^\cdot[A_\infty^2] \le 4\, E^\cdot[B_\infty^2]$$

where A and B are the additive functionals of r and s respectively. In particular,

$$\|r\|_e \le 2\, \|s\|_e \qquad \text{if } r \le s.$$

Let s be a class (D) potential of finite energy whose additive functional is A. Let $s_n = Uf_n$ where s_n increases to s as $n \to \infty$. Then for each $\alpha > 0$, $s_n^\alpha = U^\alpha f_n$ converges almost everywhere to s^α where

$$s^\alpha = E^\cdot[\int_0^\infty e^{-\alpha t}\, dA_t].$$

It is proved in [5] that

$$(s_n^\alpha, f_n) \ge \frac{\alpha}{2}\, \|s_n^\alpha\|_2^2 \, ,$$

where $\|\cdot\|_2$ is the L^2 - norm. Using (10) we conclude:

PROPOSITION 4. *Let* s *be a class* (D) *potential of finite energy. Then for each* $\alpha > 0$, s^α *is square integrable and*

$$\alpha \|s^\alpha\|_2^2 \leq 8 \|s\|_e^2 . \qquad \square$$

In the classical Brownian motion case a well-known result [3] states that, for a sequence of potentials that is bounded in energy, weak convergence -in energy sense- is equivalent to vague convergence of corresponding measures. This theorem has an analogue in the general case, which may be stated as follows:

Let s_n be a sequence of class (D) potentials that is bounded in energy norm. Suppose also that $L(s_n)$ is uniformly bounded. Then weak convergence in energy of s is equivalent to weak convergence of the corresponding Revuz measures. Proof of this statement will be given elsewhere.

Finally, let us mention a very simple fact. Let s be a class (D) potential of finite energy and let A be its additive functional. Write

$$t(\cdot) = E^{\cdot}[(A_\infty - s(X_0))^2] \tag{11}$$

and let M denote the square integrable martingale

$$M_t = E\,[A_\infty|\mathcal{F}_t]. \tag{12}$$

As we have already observed, t is in class (D). It is interesting that the increasing process of t is the increasing process of M.

Indeed, because $A_\infty = A_t + A_\infty(\theta_t)$,

$$E^x[A_\infty^2 | \mathcal{F}_t] = E^{X_t}[A_\infty^2] + 2 A_t E^{X_t}[A_\infty] + A_t^2$$

$$= E^{X_t}[A_\infty^2] + 2 A_t s(X_t) + A_t^2$$

$$= t(X_t) + s^2(X_t) + 2 A_t s(X_t) + A_t^2 = t(X_t) + M_t^2 ,$$

showing that the potential $t(X)$ and the submartingale M^2 have the same increasing process.

References

1. R.M. BLUMENTHAL and R.K. GETOOR. *Markov Processes and Potential Theory.* Academic Press, New York, 1968.
2. C. DELLACHERIE and P.A. MEYER. *Probabilités et Potentiel, Vol. II.* Hermann, Paris, 1980.
3. L.L. HELMS. *Introduction to Potential Theory.* Robert E. Krieger Publishing Company, Huntington New York, 1975.
4. P.A. MEYER. *Probability and Potentials.* Blaisdell, Waltham, 1966.
5. Z.R. POP-STOJANOVIC and MURALI RAO. Some results on Energy. *Seminar on Stochastic Processes 1981,* pp. 135-150. Birkhäuser, Boston, 1981.
6. Z.R. POP-STOJANOVIC and MURALI RAO. Remarks on Energy. *Seminar on Stochastic Processes 1982,* pp. 229-235. Birkhäuser, Boston, 1983.
7. D. REVUZ. Mesures associées à fonctionnelles additive de Markov. *Trans. Amer. Math. Soc. 148* (1970) 501-531.

Department of Mathematics
University of Florida
Gainesville, Florida 32611

Seminar on Stochastic Processes, 1983
Birkhäuser, Boston, 1984

A DIFFUSION FIRST PASSAGE PROBLEM

by

L.C.G. ROGERS

1. Introduction

Let $(X_t)_{t\geq 0}$ be a diffusion in natural scale on $I \subseteq [0,\infty)$, with speed measure m and initial distribution μ. Suppose that $0 \in I$, and 0 is a regular boundary point (without loss of generality we may suppose the diffusion is reflected instantaneously at 0). Let $\tau \equiv \inf\{t;\ X_t = 0\}$. Suppose that X' is another diffusion on I, again in natural scale and with speed measure m, but this time with initial distribution μ'. Let $\tau' \equiv \inf\{t;\ X'_t = 0\}$. The problem is this:

(1) *Suppose* τ *and* τ' *have the same distribution; does it follow that* $\mu = \mu'$?

Though this is a very simple question to state, it appears that the simplicity ends at the statement! The only proof yet known relies on a mountain of differential equations results. We shall show that, under a technical condition, the answer to (1) is "Yes". We give an example where the technical condition is not satisfied, but the answer to (1) is still "Yes", raising the conjecture that the technical condition is superfluous.

2. Spectral Theory of One Dimensional Diffusions

We shall set out some well-known results on one dimensional diffusions and the spectral theory of their generators; see Itô-McKean [2], Mandl [3], and Dym-McKean [1] for full details.

Let $\mathfrak{G}$ denote the generator $\frac{1}{2}\frac{d^2}{dmdx}$ of the diffusion X, and for each $\omega \in \mathbb{C}$, let $A(x,\omega)$ denote the solution to

$$\mathfrak{G}f = -\tfrac{1}{2}\,\omega^2 f, \quad f(0) = 1, \quad f'(0) = 0. \tag{2}$$

For the existence of A and further properties, see [1], p. 171. Notice that, since the behavior of the diffusion at 0 is irrelevant to our problem, we may suppose $m(0) = 0$, and the right and left derivatives at zero of elements of the domain of $\mathfrak{G}$ coincide. The probabilistic interpretation of A is given by

$$A(x,\ i\sqrt{2\lambda}) \equiv 1/E^0(e^{-\lambda\tau_x})$$

for $\lambda > 0$, where $\tau_x \equiv \inf\{t;\ X_t = x\}$.

The differential equation $\mathfrak{G}f = -\frac{1}{2}\,\omega^2\, f$ has a complementary solution; its construction is slightly intricate for $\omega^2 \in \mathbb{C}\backslash[0,\infty)$, but if we restrict attention to $\omega^2 \in (-\infty,0)$ the construction is straightforward. As this is the only case of interest to us, we define $D(x,ib)$ for $b \in \mathbb{R}$ to be the unique solution to

$$\mathfrak{G}f = \tfrac{1}{2}\, b^2 f, \quad f(0) = 1, \quad f \geq 0, \quad f \quad \textit{decreasing}.$$

Dym-McKean [1, p. 164] show how to construct f by differential equations techniques; probabilistically, it is defined by

$$D(x,ib) \equiv E^x[e^{-\frac{1}{2}b^2\tau_0}].$$

Thus our question (1) can be restated:

If $\int_I \mu(dx)\, D(x,ib) = \int_I \mu'(dx)\, D(x,ib)$ *for all* $b \in \mathbb{R}$, *is* $\mu = \mu'$?

Or again,

(3) *Is the family* $\{D(\cdot,ib);\ b\geq 0\}$ *total in* $C(I)$?

The answer to this question is not known, but the answer to a closely related question is known.

THEOREM 1. *The family* $\{D(\cdot,ib);\ b>0\}$ *is total in* $H \equiv L^2(I,m)$.

PROOF. There is a measure Δ on $\mathbb{R}^+$, finite for compacts, such that there is an isometry of H and $L^2(\mathbb{R}^+,\Delta)$, this isometry being given by

(4) $$f \to \hat{f}(\gamma) = \int_I A(x,\gamma)\, f(x)\, m(dx),$$

with the explicit inversion formula

(5) $$f(x) = \frac{1}{\pi}\int_{\mathbb{R}^+} A(x,\gamma)\, \hat{f}(\gamma)\, \Delta(d\gamma);$$

(see [1, p. 186]; this is the generalization of the Fourier cosine transform, with which it agrees if $m(dx) = dx$). Since the map is an isometry,

$$(f,g) = \langle \hat{f},\hat{g} \rangle$$

for all f, $g \in H$, where $(\cdot,\cdot)$ denotes the inner product in H, and $\langle\cdot,\cdot\rangle$ denotes the inner product in $L^2(\mathbb{R}^+,\Delta)$. The measure Δ also satisfies the integrability condition

$$\int_{\mathbb{R}^+} (1+x^2)^{-1}\,\Delta(dx) < \infty.$$

We shall prove that $\{D(\cdot,ib);\ b\geq 0\}$ is total in H by proving that the transforms of elements of this family are total in $L^2(\mathbb{R}^+,\Delta)$.

If for each $\lambda > 0$ we define the resolvent R_λ of the diffusion X by

$$R_\lambda f(x) = E^x[\int_0^\infty e^{-\lambda t} f(X_t)\,dt]$$

for bounded measurable f, then it is well known that $R_\lambda(x,dy)$ has a density $r_\lambda(x,y)$ with respect to the speed measure, and $r_\lambda(x,y)$ has the explicit form

$$(6) \qquad r_\lambda(x,y) = \begin{cases} p_\lambda\ A(x,ib)\ D(y,ib) & \text{if} \quad x\leq y \\ p_\lambda\ A(y,ib)\ D(x,ib) & \text{if} \quad x\geq y \end{cases}$$

where $b \equiv \sqrt{2\lambda}$, and

$$p_\lambda \equiv 2\{D(x,ib)\,A'(x,ib) - A(x,ib)\,D'(x,ib)\}^{-1},$$

which is independent of x. See [1], [2], or [3]. Now, Dym-McKean [1, p. 176] prove that the resolvent density can be expressed in terms of the eigenfunctions $A(\cdot,\gamma)$ by

$$(7) \qquad r_\lambda(x,y) = \frac{1}{\pi}\int_{\mathbb{R}^+} \frac{A(x,\gamma)\, A(y,\gamma)}{\gamma^2 + 2\lambda}\, \Delta(d\gamma).$$

Noticing that each $D(\cdot,ib)$ is in H (for they are each bounded, and integrable) we can ask what are the transforms $\hat{D}_{ib}$ of $D(\cdot,ib)$. The answer comes from inspection of (5), (6), and (7). Taking $x = 0$ in (7), the left hand side is $r_\lambda(0,y) = p_\lambda\, D(y,ib)$ and the right hand side is

$$\frac{1}{\pi}\int_{\mathbb{R}^+} \frac{A(y,\gamma)}{\gamma^2 + 2\lambda}\, \Delta(d\gamma).$$

Comparing now with (5), we deduce immediately that

$$p_\lambda\, \hat{D}_{ib}(\gamma) = \frac{1}{\gamma^2 + 2\lambda}.$$

Thus $\{\hat{D}_{ib};\, b>0\}$ is total in $L^2(\mathbb{R}^+,\Delta)$ (the Stieltjes transform determines the measure!) and hence $\{D(\cdot,ib);\, b>0\}$ is total in H, as required.

3. The Main Result

Let R^∂_λ denote the resolvent of X killed when it first hits zero, with r^∂_λ the resolvent density; if $\psi_\lambda(x) \equiv E^x[e^{-\lambda\tau_0}] \equiv D(x,i\sqrt{2\lambda})$ and $\varphi_\lambda(x) \equiv A(x,i\sqrt{2\lambda}) - D(x,i\sqrt{2\lambda})$, then

$$r^\partial_\lambda(x,y) = \begin{cases} p_\lambda\, \varphi_\lambda(x)\, \psi_\lambda(y) & (y \geq x) \\ \\ p_\lambda\, \varphi_\lambda(y)\, \psi_\lambda(x) & (y \leq x) \end{cases}$$

where p_λ is as defined before. Here then is our main result.

THEOREM 2. *Suppose that* μ, μ' *are two probability measures on* I *such that for each* $\lambda > 0$

$$\int_I \mu(dx)\, E^x[e^{-\lambda\tau_0}] = \int_I \mu'(dx)\, E^x[e^{-\lambda\tau_0}]. \tag{8}$$

If, further, the condition

$$\sup_{x,y}\{\int_I r_\lambda^\partial(x,z)\, r_\lambda^\partial(z,y)\, m(dz)\} < \infty \quad \textit{for each} \quad \lambda > 0 \tag{U}$$

is satisfied, then $\mu = \mu'$.

REMARK. U stands for unnatural; it is conjectured that it also stands for unnecessary!

PROOF. For all $\lambda, \alpha > 0$,

$$\int_I (\mu-\mu')(dx)\, R_\alpha^\partial \psi_\lambda(x) = 0;$$

this is obvious to a probabilist, and an analyst will see that it is true from the fact that $\psi_\lambda = 1 - \lambda R_\lambda^\partial 1$, using the resolvent equation to give $R_\alpha^\partial \psi_\lambda = (\alpha-\lambda)^{-1}(\psi_\lambda - \psi_\alpha)$. Re-expressing this,

$$\int_I \{\int_I (\mu-\mu')(dx)\, r_\alpha^\partial(x,y)\}\, \psi_\lambda(y)\, m(dy) = 0\ ; \tag{9}$$

that is, we may replace μ by μR_α^∂ and suppose that μ (and likewise μ') has a density with respect to m. <u>Provided</u> this density is square integrable, we may use the fact that $\{\psi_\lambda;\ \lambda>0\}$ is total in H (Theorem 1) to deduce that

$$\int \mu(dy)\, r_\alpha^\partial(y,x) = \int \mu'(dy)\, r_\alpha^\partial(y,x)$$

for all x, $\alpha > 0$; letting $\alpha \to \infty$ we conclude $\mu = \mu'$.

As for the square integrability of the density of μR_α^∂,

$$\int m(dx)\, \left(\int \mu(dy)\, r_\alpha^\partial(y,x)\right)^2 = \int \mu(dy) \int \mu(dz) \int m(dx)\, r_\alpha^\partial(y,x)\, r_\alpha^\partial(x,z)$$

by the symmetry of $r_\alpha^\partial(\cdot,\cdot)$; and this is finite by condition U.

REMARK. Using the fact that

$$\int r_\lambda^\partial(x,z)\, r_\lambda^\partial(z,y)\, m(dz) = -\frac{\partial}{\partial\lambda}\, r_\lambda^\partial(x,y)$$

we can often make a check of condition U more easily.

4. An Example

Suppose $(Y_t)_{t\geq 0}$ is a Brownian motion with drift $-\delta < 0$. Then for $x > 0$

$$E^x[e^{-\lambda\tau_0}] = \exp\{-((\delta^2+2\lambda)^{\frac{1}{2}} - \delta)x\}$$

and so if μ, μ' are two measures on $(0,\infty)$ such that for all $\lambda > 0$

$$\int \mu(dx)\, E^x[e^{-\lambda\tau_0}] = \int \mu'(dx)\, E^x[e^{-\lambda\tau_0}]$$

it is immediate that $\mu = \mu'$. Thus the conclusion of Theorem 2 holds for this diffusion whether or not condition U is satisfied. We shall show that condition U is <u>not</u> satisfied, though we have to transform the problem to natural scale first to put it into the form previously studied. The scale function of Y is $s(y) = e^{2\delta y}$, so if $X_t = s(Y_t)$, then X is a diffusion in natural scale on $(0,\infty)$ with generator

$$\mathbf{G} = 2\delta^2 x^2 \frac{d^2}{dx^2} .$$

We only consider X in $[1,\infty)$, killing when it reaches 1. The speed measure is $m(dy) = (2\delta y)^{-2}\, dy$, and a few calculations show that the linearly independent eigenfunctions are

$$\psi_\lambda(x) = x^{-c_-(\lambda)} \qquad (x\geq 1)$$

$$\boldsymbol{\varphi}_\lambda(x) = x^{c_+(\lambda)} - x^{-c_-(\lambda)} \qquad (x\geq 1)$$

where

$$2\delta\, c_+(\lambda) \equiv (\delta^2 + 2\lambda)^{\frac{1}{2}} + \delta$$

$$2\delta\, c_-(\lambda) \equiv (\delta^2 + 2\lambda)^{\frac{1}{2}} - \delta.$$

Thus the density of the killed resolvent is

$$r_\lambda^\partial(x,y) = p_\lambda\, \boldsymbol{\varphi}_\lambda(x)\, \psi_\lambda(y) \qquad (1\leq x\leq y)$$

where, as before,

$$p_\lambda \equiv 2\{\psi_\lambda(x)\, \boldsymbol{\varphi}_\lambda'(x) - \boldsymbol{\varphi}_\lambda(x)\, \psi_\lambda'(x)\}^{-1} = 2\delta(\delta^2 + 2\lambda)^{-\frac{1}{2}}.$$

So to check condition U, by the remark at the end of the previous section, we must consider $-\frac{\partial}{\partial\lambda} r_\lambda^\partial(x,y)$. Now for $1 \leq x \leq y$

$$-\frac{1}{2\delta}\frac{\partial}{\partial\lambda} r_\lambda^\partial(x,y) = \frac{\varphi_\lambda(x)\,\psi_\lambda(y)}{(\delta^2+2\lambda)^{3/2}} - \frac{1}{2\delta(\delta^2+2\lambda)}\{\log(x/y)\, y^{-c_-(\lambda)}\, x^{c_+(\lambda)}$$

$$+ \log(xy)\,(xy)^{-c_-(\lambda)}\}.$$

Thus if we take $x = y$ we have

$$-\frac{1}{2\delta}\frac{\partial}{\partial\lambda} r_\lambda^\partial(x,x) = \frac{x - x^{-2c_-(\lambda)}}{(\delta^2+2\lambda)^{3/2}} - \frac{\log x}{\delta(\delta^2+2\lambda)}\, x^{-2c_-(\lambda)}$$

$$\sim \frac{x}{(\delta^2+2\lambda)^{3/2}} \quad \textit{as} \quad x \to \infty.$$

Thus condition U is violated, even though the conclusion of Theorem 2 remains valid.

As we have seen, condition U only comes in because we are using an indirect approach to the problem; we want to prove that $\{\psi_\lambda;\ \lambda>0\}$ is total in $C(I)$ but can only prove it total in $L^2(I,m)$. The techniques for proving totality in $C(I)$ appear not to have been developed. This is suprising firstly because such problems arise frequently in probability, and secondly because such results as there are (Stone-Weierstrass, Laplace and Fourier transforms determine measures) are only applicable in special circumstances. The whole area calls out for a systematic study!

ACKNOWLEDGEMENT. I should like to thank all of those at the conference with whom I discussed this problem for their helpful comments.

References

[1] H. DYM and H.P. McKEAN. *Gaussian Processes, Function Theory, and the Inverse Spectral Problem*. Academic Press, New York, 1976.

[2] K. ITÔ and H.P. McKEAN. *Diffusion Processes and Their Sample Paths*. Springer-Verlag, Berlin, 1965.

[3] P. MANDL. *Analytic Treatment of One-Dimensional Markov Processes*. Springer-Verlag, Berlin, 1968.

L.C.G. ROGERS
Department of Mathematics
University College of Swansea
Swansea SA2 8PP
Great Britain

Seminar on Stochastic Processes, 1983
Birkhäuser, Boston, 1984

BROWNIAN EXCURSIONS REVISITED*

by

P. SALMINEN

1. Introduction

There are two classical approaches to the theory of Brownian excursions. The first one goes back to Lévy. His ideas were worked out in greater detail and extended by Itô and McKean (see [4], [5], and [9]). Also Chung's and Knight's contributions are of great importance (see [1], [7], and [8]). In this approach the lengths of the excursions are the basic objects. In the second approach, due to Williams (see [12], [14], and [15]), one works with excursions having a given maximum. In both approaches Itô's theory of excursions (see [3]) plays an active part (see [5], and [12]).

We focus to Lévy's approach, but our descriptions are in the spirit of Itô and Williams. However, the aim of the paper is to demonstrate the usefullness of both of these approaches. Especially, in the fifth section we show how to derive more intuitive descriptions of the excursion law from Lévy's approach.

---

*Research sponsored by Magnus Ehnrooth Foundation, Finland, and by the Air Force Office of Scientific Research, under grant number AFOSR-82-0189.

As a preliminary, in the next section, we show how to construct the so-called diffusion bridges in the framework of the Martin boundary theory. This kind of construction seems to be new.

The third section is devoted to a proof of the Itô excursion law of the reflected Brownian motion. Our proof is somewhat shorter than Knight's presented in [7].

In the fourth section we calculate the distribution of the maximum of a three dimensional Besselian bridge. It is seen that this contains many essential points of the theory of Brownian excursions. Further, our formula is extended to give the distribution of the difference of the last exit time from a point x and the first hitting time of x in a three dimensional Besselian bridge with the maximum greater than x. It is seen that this distribution also splits naturally into components which all have a clear explanation. These results should be compared with the results in [8].

In [1] and [2], for example, the excursions straddling a fixed time were investigated. In [2] Getoor and Sharpe establish limit theorems for sojourn and local times. We show how to obtain these theorems rather easily using Williams' characterization of the excursion law.

To apply the approach based on the lengths of excursions one has to have, of course, the distribution of the lengths. Motivated by this we conclude by offering a new proof for a construction of the Lévy measure for a Lévy process which is the right-continuous inverse of the local time of a diffusion.

The underlying sample space Ω is the set of all functions ω from $[0,\infty]$ to $(-\infty,+\infty) \cup \{\Delta\}$ (Δ is some fictitious state) such that, for some $T(\omega) \in (0,+\infty]$, ω is continuous on $[0,T(\omega))$ and takes the value Δ on $[T(\omega),+\infty]$. We let (X_t) be the coordinate process on Ω and let $F = \sigma(X_s;\ s\geqq 0)$ and $F_t = \sigma(X_s;\ s\leqq t)$.

2. Diffusion Bridges

We give our construction for a general, regular, canonically defined one-dimensional diffusion. Let $(\mathbb{P}_x)_{x\in(a,b)}$ be a family of probability measures so that $X = (X_t, \mathbb{P}_x)$ is such a diffusion taking values in an interval $I = (a,b)$. If a (resp. b) is not natural we assume a (resp. b) $\in I$. Further, we assume that there is no killing inside I. If a boundary point is entrance-exit, the boundary condition for X can be instantaneous reflection or killing.

It is well-known that X has a smooth transition density $p(t; x,y)$ with respect to the speed measure m of X (for its properties see [4] p. 149). In particular, $(t,x,y) \to p(t; x,y)$ is continuous and symmetric in (x,y): $p(t; x,y) = p(t; y,x)$.

Consider X in space-time. For $\ell > 0$ and $z \in I$, let

(1)
$$h_{\ell,z}(s,y) = \begin{cases} p(\ell-s;\, y,z), & \text{if } s < \ell, \\ 0, & \text{otherwise.} \end{cases}$$

It follows from Chapman-Kolmogorov equation that $h_{\ell,z}$ is a space-time excessive function for $X = (X_t, \mathbb{P}_x)$. Let $\mathbb{P}^{\ell}_{x,z}$ be the probability law of the $h_{\ell,z}$-transform or X.

(2) REMARK. If z in (1) is a boundary point then it is possible that $p(\ell-s;\, y,z) = 0$. In this case we set

(3)
$$h_{\ell,z}(s,y) = \begin{cases} \lim\limits_{\alpha\to z} \dfrac{p(\ell-s;\, y,\alpha)}{p(\ell;\, x,\alpha)}, & \text{if } s < \ell, \\ 0, & \text{otherwise;} \end{cases}$$

here x is the starting state of X.

(4) PROPOSITION. For all $x \in I$ ($x \neq a$ (resp. b) if a (resp. b) is exit, non-entrance, $\lim\limits_{t\to\ell} X_t(\omega) = z$ for $\mathbb{P}^{\ell}_{x,z}$-almost every ω.

PROOF. It is clear that $\lim_{t\to\ell} X_t(\omega)$ exists for all ω. Let f be a positive, bounded, and continuous function that vanishes over a neighborhood of z. We have

$$\mathbb{P}^{\ell}_{x,z}(f\circ X_t) = \int_I \frac{1}{p(\ell;\ x,z)}\, p(t;\ x,y)\ f(y)\ p(\ell-t;\ y,z)\ m(dy)$$

$$\leq C\ \mathbb{P}_z(f\circ X_{\ell-t})$$

for some constant C not depending on t. Here we used both the continuity and the symmetry of $p(t;\ x,y)$.

Since f vanishes over a neighborhood of z, $f\circ X_{\ell-t} \to f(X_0) = f(z) = 0$ as $t \to \ell$ almost surely $(\mathbb{P}_z)$ by the continuity of f and X. So, by the bounded convergence theorem, $\mathbb{P}_z(f\circ X_{\ell-t}) \to 0$ as $t \to \ell$, and hence $\mathbb{P}^{\ell}_{x,z}(f\circ X_t) \to 0$ as $t \to \ell$ in view of the inequality above. This implies the proposition.

We call $\mathbb{P}^{\ell}_{x,z}$ the law of X bridge of length ℓ from x to z.

(5) REMARKS. (i) From the construction it is clear that $\mathbb{P}^{\ell}_{x,z}$ could also be described as the regular conditional $\mathbb{P}_x$-distribution of $(X_s;\ 0\leq s\leq\ell)$ given $\{X(\ell) = z\}$.

(ii) Assume that the boundary point a is exit-entrance and the boundary condition is killing. Further assume that $X_{\zeta-} = a$ $\mathbb{P}_x$-a.s. Then, $p(t;\ x,a) = 0$ for all x and so (see [4] p. 154)

$$\mathbb{P}_x(\tau_a\in dt) = \lim_{y\downarrow a} \frac{p(t;\ x,y)}{s(y)-s(a)}\ dt.$$

It follows that $\mathbb{P}^{\ell}_{x,a}$ can be interpreted as the regular conditional $\mathbb{P}_x$-distribution of X given $\{\tau_a=\ell\}$.

(iii) Assume that X is transient and that $s(a) = -\infty$ and $s(b) < +\infty$. Then $\lim_{t\uparrow\zeta} X_t = b$ $\mathbb{P}_x$-a.s. Let $\lambda_y = \sup\{t\colon X_t=y\}$. We have, for $x < y$ (see [11] Theorem 6.1 or [13]),

$$\mathbb{P}_x(\lambda_y \in dy) = \frac{p(t;\ x,y)}{s(b)-s(y)}\, dt.$$

Consequently $\mathbb{P}^{\ell}_{x,y}$ is the regular conditional $\mathbb{P}_x$-distribution X given $\{\lambda_y = \ell\}$.

Because X bridge is a space-time h-transform of X and X is a strong Markov process, it follows that the X bridge has also the strong Markov property at stopping times $\tau < \ell$. Moreover it is immediate from our construction that

$$\mathbb{P}^{\ell}_{x,z}\,\Big|_{F_\tau} << \mathbb{P}_x\,\Big|_{F_\tau}$$

and

(6) $$\frac{d\ \mathbb{P}^{\ell}_{x,z}}{d\ \mathbb{P}_x}\Bigg|_{F_\tau} = \frac{p(\ell-\tau;\ X_\tau, z)}{p(\ell;\ x,z)}\ .$$

Next, let $X = (X_t, \mathbb{R}_0)$ be a three dimensional Bessel process started from 0, denoted BES_0, i.e. a diffusion on $(0,+\infty)$ governed by the differential operator

$$A = \frac{1}{2}\, D^2 + \frac{1}{x}\, D,$$

where $D = \frac{d}{dx}$. We can take $s(x) = -\frac{1}{x}$ and $m(dx) = 2x^2 dx$ as a scale function and a speed measure, respectively. Further, it is easy to see that 0 is entrance non-exit boundary point, and that $+\infty$ is natural.

We are interested in X bridge from 0 to 0 with a given length ℓ. This process is denoted $BESB^{\ell}_{00}$ and its law with $\mathbb{R}^{\ell}_{00}$. In this case (6) takes the form

$$(6') \qquad \left.\frac{d\,\mathbb{R}^{\ell}_{00}}{d\,\mathbb{R}_0}\right|_{F_\tau} = \left(\frac{\ell}{\ell-\tau}\right)^{3/2} \exp\left(-\frac{1}{2(\ell-\tau)}\, X_\tau^2\right).$$

From the absolute continuity and the fact that

$$\mathbb{R}_0(X_t=0 \quad \text{for some} \quad 0<t<\ell) = 0$$

it follows immediately that

$$(7) \qquad \mathbb{R}^{\ell}_{00}(X_t=0 \quad \text{for some} \quad 0<t<\ell) = 0.$$

(8) REMARK. In [5] pp. 225-229, Ikeda and Watanabe prove the existence of $BESB^{\ell}_{00}$ by constructing stochastic differential equation and showing that the unique solution of this equation has the law $\mathbb{R}^{\ell}_{00}$. They also give an argument for (7).

Let $\hat{X} = (X_t, \hat{\mathbb{P}}_x)$ be a Brownian motion started from x and killed when it hits 0, for short ABM_x. Denote with $\hat{\mathbb{P}}^{\ell}_{x,0}$ the law of the $\hat{X}$ bridge of length ℓ from x to 0; this bridge is denoted $ABMB^{\ell}_{x0}$. Then we have:

(9) The laws $\hat{\mathbb{P}}^{\ell}_{x,0}$ and $\mathbb{R}^{\ell}_{x,0}$ are the same.

This fact, in slightly weaker form, goes back to McKean (see [9]). To prove this is straightforward and we omit it (however, see [7] for a proof of the weaker form).

(10) REMARK. By Remark (5, ii) the law $\hat{\mathbb{P}}^{\ell}_{x,0}$ is the regular conditional $\hat{\mathbb{P}}_x$-distribution of X given $(\tau_0=\ell)$. Consequently, we have

(11) $$\hat{\mathbb{P}}_x(\cdot) = \int_0^\infty \hat{\mathbb{P}}^{\ell}_{x,0}(\cdot)\ \hat{\mathbb{P}}_x(\tau_0 \in d\ell).$$

A third bridge of interest for our further developments is constructed from the diffusion on $[0,\infty)$ having the generator

$$\mathcal{B} = 2xD^2 + 2D.$$

We can take $s(y) = \log y$ and $m(dy) = \frac{1}{2}\,dy$ as a scale function and a speed measure, respectively. In fact, the diffusion having the generator $\mathcal{B}$ is the square of a two-dimensional Bessel process. The boundary point 0 is entrance, not-exit and $+\infty$ is natural. Transition densities (with respect to m) are given by

$$p(t;\ x,y) = \frac{1}{t}\exp\left(-\frac{x+y}{2t}\right) I_0\left(\frac{\sqrt{xy}}{t}\right),$$

where I_0 is the modified Bessel function of the first kind and of order 0. Consider a bridge of length ℓ from 0 to 0. Because $I_0(x) \to 1$ as $x \to 0$, transition densities are given by

$$p^b(t;\ x,y) = p(t;\ x,y)\,\frac{g(t,y)}{g(0,x)},$$

where

$$g(t,y) = \frac{\ell}{\ell - t}\exp\left(-\frac{y^2}{2(\ell-t)}\right).$$

Especially

(12) $$p^b(t;\ 0,y)\ m(dy) = \frac{\ell}{2t(\ell-t)}\exp\left(-\frac{\ell}{2t(\ell-t)}\,y\right) dy.$$

See [10] for properties, decompositions and interpretations of Besselian bridges.

3. Itô Excursion Law of the Reflected Brownian Motion

Let $X = (X_t, \mathbb{P}_x)$ be a Brownian motion reflected at 0 and started from x; for short RBM_x. We shall describe briefly the Itô excursion law of X for excursions away from 0.

Let U be the set of all continuous functions $f: [0,+\infty) \to [0,+\infty)$ such that $f(0) = 0$, $f(t) > 0$ for $t \in (0,\zeta_f)$ and $f(t) = 0$ for $t \geq \zeta_f$, for some $\zeta_f > 0$. We give U the topology of uniform convergence on compacts, and let $\mathcal{U}$ be the corresponding Borel field.

Let $b > 0$ be given and denote with η the time point (a stopping time), when the first excursion of the length $\geq b$ ends. Further let $\sigma = \sup\{t<\eta: X(t) = 0\}$ and introduce $e(t) = X(\sigma+t)$, $0 \leq t < \eta - \sigma$.

The following result is well-known; for a proof see, for example, [7] theorem 5.2.7. Notice, however, that our formulation leads immediately to Theorem 3 below, also our proof is perhaps somewhat simpler than the one in [7].

(1) PROPOSITION. The process $X^e = \{e(t);\ 0\leq t\leq \eta-\sigma\}$ is identical in law with $BESB^{\ell}_{00}$, where $\ell = \eta - \sigma$.

PROOF. Clearly, the lengths of the excursions are governed by the σ-finite measure

$$n(d\ell) = \frac{1}{\sqrt{2\pi\ell^3}}\, d\ell .$$

Denote with $\mathbb{P}^e$ the law of X^e and let α be a positive and bounded random variable measurable with respect to the σ-field generated by $X(t)$ with $\sigma \leqq t < \eta$. Then we have to prove

$$\mathbb{P}^{e}(\alpha) = \int_{b}^{\infty} \mathbb{R}_{00}^{\ell}(\alpha)\ n(d\ell \mid \ell \geqq b). \tag{2}$$

Of course, $n(\cdot \mid \ell \geqq b)$ is a probability measure defined by

$$n(\cdot \mid \ell \geqq b) = \frac{n(\cdot;\ \ell \geqq b)}{n(\ell \geqq b)}$$

Now, given $\varepsilon > 0$ let X^{ε} be a process which starts from ε and moves like a Brownian motion until it hits 0. Then it jumps to ε and repeats itself. The claim is that as $\varepsilon \downarrow 0$ the law of X^{ε} is just the law of RBM_0.

To see this let Y be BM_0 and $M_t = \sup_{s \leqq t} Y_t$. Consider the process Z^{ε} defined as follows

$$Z_t^{\varepsilon} = \left(\left[\frac{M_t}{\varepsilon}\right] + 1\right)\varepsilon - Y_t, \tag{3}$$

where $[x]$ is the integer part of x. It is easily seen that Z^{ε} and X^{ε} are identical in law. But as $\varepsilon \downarrow 0$ in (3) we get

$$Z_t^0 = M_t - Y_t.$$

This shows that Z^0 is identical in law with RBM_0. Consequently, the law of X^{ε} converges (weakly) as $\varepsilon \downarrow 0$ to the law of RBM_0.

Let $\mathbb{P}_{\varepsilon}^{e}$ be the law of the first X^{ε} excursion with the length $\geqq b$. Then it is clear from above that as $\varepsilon \downarrow 0$ $\mathbb{P}_{\varepsilon}^{e}$ converges (weakly) to $\mathbb{P}^{e}$. Further, let $\tau_0 = \inf\{t: X(t) = 0\}$ then

$$\mathbb{P}_{\varepsilon}^{e}(\alpha) = \int_{b}^{\infty} \hat{\mathbb{P}}_{\varepsilon 0}^{\ell}(\alpha)\ \hat{\mathbb{P}}_{\varepsilon}(\tau_0 \in d\ell \mid \tau_0 \geqq b). \tag{4}$$

where $\hat{\mathbb{P}}_\varepsilon$ and $\hat{\mathbb{P}}^\ell_{\varepsilon 0}$ are the laws of ABM_ε and $ABMB^\ell_{\varepsilon 0}$, respectively. It is easily seen that $\hat{\mathbb{P}}_\varepsilon(\tau_0 \ \bullet \mid \tau_0 \geq b)$ converges (weakly) to $n(\bullet \mid \ell \geq b)$ as $\varepsilon \downarrow 0$. But by (2.9) $\hat{\mathbb{P}}^\ell_{\varepsilon 0} = \mathbb{R}^\ell_{\varepsilon 0}$. Hence, letting $\varepsilon \downarrow 0$ in (4) gives us (2), and the proof is complete.

Now we can follow Rogers (see [12] p. 235, and Proposition 3.2) to give a proof of

(5) THEOREM. The Itô excursion law of RBM_0 is the σ-finite measure ν on $(U, \mathcal{U})$ defined by

$$\nu(F) = \int_0^\infty n(d\ell)\ \mathbb{R}^\ell_{00}(F), \qquad F \in \mathcal{U}.$$

Notice that $\mathbb{R}^\ell_{00}(F) = \mathbb{R}^\ell_{00}(F \cup \{\zeta=\ell\})$ and, hence,

$$\nu(\zeta>\ell) = n([\ell,+\infty)) = \sqrt{\frac{2}{\pi\ell}}, \qquad \ell > 0,$$

which shows that ν is σ-finite.

By the theory of Itô there exists a Poisson random measure on $[0,+\infty) \times U$ with mean measure $dt \times \nu(df)$. Here $[0,+\infty)$ should be interpreted as the local time axis. For details and further results see [3], [4], [12], [14], and [15]. For a converse to Theorem 4 see [5] pp. 125-129.

4. An Excursion Lemma and Some Consequences

We recall some notation. For the canonical process X let τ_x be the hitting time of x and M_t the maximum during $[0,t]$. Let $\mathbb{P}_x$, $\mathbb{R}_0$ and $\mathbb{R}^\ell_{00}$ be the laws, respectively, of BM_x, BES_0 and $BESB^\ell_{00}$. We introduce

$$\mathbb{P}_x(\tau_0 \in ds) \equiv b_x(s)\ ds, \qquad \mathbb{R}_0(\tau_x \in ds) \equiv r_x(s)\ ds.$$

(1) LEMMA. For $x > 0$ and $t < \ell$

$$\mathbb{R}^{\ell}_{00}(\tau_x<t) = \frac{\sqrt{2\pi\ell}^{\,3}}{x}\int_0^t r_x(s)\, b_x(\ell-s)\, ds.$$

PROOF. Let $\hat{\tau}_x = \tau_x \wedge t$. Then $\hat{\tau}_x < \ell$ and $\{\tau_x<t\} = \{\hat{\tau}_x<t\} \in F_{\hat{\tau}_x}$. Therefore using (2.6′)

$$\mathbb{R}^{\ell}_{00}(\tau_x<t) = \mathbb{R}^{\ell}_{00}(\hat{\tau}_x<t) = \mathbb{R}_0(\hat{\tau}_x<t;\ h(\hat{\tau}_x, X_{\hat{\tau}_x})),$$

where (cf. (2.6′))

$$h(s,x) = \left(\frac{\ell}{\ell-s}\right)^{3/2} \exp\left(-\frac{1}{2(\ell-s)}\, x^2\right). \tag{2}$$

But on $\{\hat{\tau}_x<t\}$ $\tau_x \equiv \hat{\tau}_x$ and hence we obtain

$$\begin{aligned}\mathbb{R}^{\ell}_{00}(\tau_x<t) &= \int_0^t \mathbb{R}_0(\tau_x\in ds)\ h(s,x)\\ &= \frac{\sqrt{2\pi\ell}^{\,3}}{x}\int_0^t r_x(s)\, b_x(\ell-s)\, ds.\end{aligned}$$

(3) COROLLARY. For $x > 0$

$$\mathbb{R}^{\ell}_{00}(M_{\ell}>x) = \frac{\sqrt{2\pi\ell}^{\,3}}{x}\int_0^{\ell} r_x(s)\, b_x(\ell-s)\, ds.$$

PROOF is immediate from the preceding lemma; just let $t \uparrow \ell$ after noting that $\{M_t>x\} = \{\tau_x<t\}$.

(4) REMARKS. (i) By a well-known time reversal argument

$$\mathbb{P}_x(\tau_0\in ds) \equiv \mathbb{R}_0(\lambda_x\in ds),$$

where $\lambda_x = \sup\{t: X(t) = x\}$. Then

$$\mathbb{R}^{\ell}_{00}(M_{\ell}>x) = \frac{\sqrt{2\pi\ell}^3}{x}\int_0^{\ell} \mathbb{R}_0(\lambda_x\in ds)\ r_x(\ell-s).$$

(ii) The basic computation can be presented as an infinite sum

$$\mathbb{R}^{\ell}_{00}(M_{\ell}>x) = 2\sum_{k=1}^{\infty}\left(4\frac{k^2x^2}{\ell} - 1\right)\exp\left(-2\frac{k^2x^2}{\ell}\right).$$

(see [1] Theorem 4, [8], and [15] p. 99).

Let $m(f) = \sup_t f(t)$, $f \in U$, and recall that ν is the excursion law of RBM_0.

(5) PROPOSITION.

a) $\nu(m \geqq x,\ \zeta \in d\ell) = \frac{1}{x}\,d\ell\int_0^{\ell} r_x(s)\ b_x(\ell-s)\ ds$

b) $\nu(m \geqq x) = \frac{1}{x}$

c) $\nu(e^{-\alpha\zeta};\ m \in dx) = dx\,\frac{2}{\sinh^2\sqrt{2\alpha}\ x}$

d) $\nu(\zeta \in d\ell \mid m \geqq x) = d\ell\int_0^{\ell} r_x(s)\ b_x(\ell-s)\ ds$

e) $\nu(\zeta \in d\ell \mid m = x) = d\ell\int_0^{\ell} r_x(s)\ r_x(\ell-s)\ ds$

f) $\nu(m \in dx \mid \zeta = \ell) = dx\,\frac{\sqrt{2\pi\ell}^3}{x^2}\int_0^{\ell} r_x(s)\ r_x(\ell-s)\ ds.$

PROOF. From the construction of ν (see Theorem 3.5) it is clear that

$$\nu(m\geqq x \mid \zeta=\ell) = \mathbb{R}^{\ell}_{00}(M_{\ell}\geqq x)$$

and

$$\nu(\zeta\in d\ell) = \frac{1}{\sqrt{2\pi\ell^3}}\, d\ell.$$

Putting these two together and using Corollary 3 yields (a). Integrating (a) over ℓ gives (b). Taking Laplace transforms in (a) we get

$$\int_0^\infty e^{-\alpha\ell}\, \nu(m\geqq x,\ \zeta\in d\ell) = \frac{1}{x}\, e^{-\sqrt{2\alpha}\, x}\, \frac{\sqrt{2\alpha}\, x}{\sinh\sqrt{2\alpha}\, x}\,.$$

Differentiation with respect to x and changing the sign gives (c). Putting (a) and (b) together gives (d). Finally, (e) follows from (b) and (c), and (f) from (e).

Next we study the joint $\mathbb{P}^{\ell}_{00}$-distribution of τ_x and λ_x. Notice that the law of Brownian motion is unchanged if all its excursions are reversed in time. This implies that the time reversal of $BESB^{\ell}_{00}$ is again $BESB^{\ell}_{00}$. In particular

$$\mathbb{P}^{\ell}_{00}(\tau_x<t) = \mathbb{P}^{\ell}_{00}(\lambda_x>\ell-t).$$

(6) PROPOSITION. For $0 < u < s < \ell$

$$\mathbb{P}_{00}(\lambda_x>s \mid \tau_x=u) = \frac{1}{2xb_x(\ell-u)} \int_0^{\ell-s} r_x(t)\, \hat{p}(\ell-u-t;\ x,x)\, dt,$$

where

$$\hat{p}(t;\ x,y) = \frac{1}{\sqrt{2\pi t}}\left(\exp\left(-\frac{(x-y)^2}{2t}\right) - \exp\left(-\frac{(x+y)^2}{2t}\right)\right).$$

PROOF. By using the strong Markov property

$$\mathbb{P}^{\ell}_{00}(\lambda_x > s \mid \tau_x = u) = \mathbb{P}^{\ell-u}_{x0}(\lambda_x > s-u).$$

Further, by the time reversal property described above and Remark 2.5 (iii) we have

$$\mathbb{P}^{\ell-u}_{x0}(\lambda_x > s-u) = \mathbb{P}^{\ell-u}_{0x}(\tau_x < \ell-s)$$

$$= \int_0^{\ell-s} \mathbb{P}_0(\tau_x \in dt)\; k(t,x),$$

where we have used the same kind of calculations as in the proof of Lemma 1. Here

$$k(t,z) = \frac{r(\ell-u-t;\; z,x)}{r(\ell-u;\; 0,x)}$$

and r is the transition density for BES (with respect to the speed measure) i.e.

$$r(t;\; y,x) = \begin{cases} \frac{1}{2}\frac{1}{xy}\,\hat{p}(t;\; x,y), & y,x \neq 0, \\ \frac{1}{\sqrt{2\pi t^3}}\exp(-\frac{x^2}{2t}), & y = 0, \end{cases}$$

and it is seen that the desired formula is obtained.

We have the following

(7) COROLLARY. For $s < \ell$

(8) $$\mathbb{P}^{\ell}_{00}(\lambda_x - \tau_x > s) = \frac{\sqrt{2\pi\ell^3}}{x}\int_s^{\ell} dt\, \frac{\hat{p}(t;\; x,x)}{2x}\int_0^{\ell-t} du\; r_x(u)\; r_x(\ell-t-u).$$

To prove this combine the results in Lemma 1 and Proposition 6.

Differentiate in (8) to obtain

$$\mathbb{P}_{00}^{\ell}(\lambda_x - \tau_x \in ds) = ds \frac{\sqrt{2\pi\ell^3}}{x} \frac{\hat{p}(s;\, x,x)}{2x} \int_0^{\ell-s} r_x(u)\, r_x(\ell-s-u)\, du.$$

To explain the terms in this formula we give

(9) PROPOSITION. For $f \in U$, let

$$\tau_x(f) = \inf\{t\colon f(t) = x\}, \quad \lambda_x(f) = \sup\{t\colon f(t) = x\}.$$

Then

a) $\nu\{\lambda_x - \tau_x \in ds,\ m \geqq x,\ \zeta \in d\ell\}$

$$= ds\,d\ell\, \frac{1}{x} \frac{\hat{p}(s;\, x,x)}{2x} \int_0^{\ell-s} r_x(u)\, r_x(\ell-s-u)\, du$$

b) $\nu\{\lambda_x - \tau_x \in ds,\ m \geqq x\} = ds\, \frac{1}{x} \frac{\hat{p}(s;\, x,x)}{2x}$

c) $\nu\{\lambda_x - \tau_x \in ds \mid m \geqq x\} = ds\, \frac{\hat{p}(s;\, x,x)}{2x}$

d) $\nu\{\zeta \in d\ell \mid \lambda_x - \tau_x = s\} = d\ell \int_0^{\ell-s} r_x(u)\, r_x(\ell-s-u)\, du, \ \ell > s.$

PROOFS are immediate after observing that

$$\mathbb{P}_{00}^{\ell}(\lambda_x - \tau_x \in ds) = \nu\{\lambda_x - \tau_x \in ds,\ m \geq x \mid \zeta = \ell\}.$$

Notice that we obtain (5.a) by integrating (9.a) with respect to s and using Proposition 6. Further (9.c) follows also from Corollary 5.2 and the formula for the last exit time distribution given in Remark 2.5 (iii).

Our answer to Knight's puzzle (see [8] p. 81) is (10.d). It seems to us that if we use (5.f) to rewrite (8) the probabilistic structure of (8) is lost. Of course, $\mathbf{P}^{\ell}_{00}(\lambda_x - \tau_x > 0) \equiv \mathbf{P}^{\ell}_{00}(M_{\ell} > x)$ and, hence, the distribution of the maximum can be deduced from (8).

5. Further Descriptions of the Excursion Law

We shall now rederive from the basic Theorem 3.5 more intuitive descriptions for the excursion law of the reflected Brownian motion.

Let X be RBM_0 and for a given $x > 0$ introduce

$$\eta_x = \inf\{t > \tau_x : X(t) = 0\},$$

$$\sigma_x = \sup\{t < \tau_x : X(t) = 0\},$$

where τ_x is the first hitting time of x. Then $X^e = \{X(t);\, \sigma_x \leq t \leq \eta_x\}$ is the first excursion with the maximum $\geq x$. We prove

(1) PROPOSITION. The process X^e is identical in law with the process Z_1 defined by

$$Z_1(t) = \begin{cases} R(t), & t \leq \tau_x \\ \hat{X}(t), & \tau_x < t, \end{cases}$$

where R is BES_0 and $\hat{X}$ is ABM_x independent of R.

PROOF. Denote the law of X^e with $\mathbf{P}^e$ and let α and β be two positive and bounded random variables measurable with respect to the σ-fields $\sigma\{X(\sigma_x + t) : 0 \leq t < \tau_x\}$ and $\sigma\{X(\tau_x + t) : 0 \leq t < \eta_x - \tau_x\}$, respectively. We have

$$\mathbb{P}^e(\alpha\beta) = \nu\{\alpha\beta \mid m \geq x\}$$

$$= \frac{\nu\{\alpha\beta;\ m \geq x\}}{\nu\{m \geq x\}}$$

$$= x \int_0^\infty \mathbb{R}_{00}^{\ell}(\alpha\beta;\ M_\ell \geq x)\ \frac{1}{\sqrt{2\pi\ell^3}}\ d\ell,$$

by Theorem 3.5. Next consider $(\tau \equiv \tau_x)$

$$\mathbb{R}_{00}^{\ell}(\alpha\beta;\ M_\ell \geq x) = \mathbb{R}_{00}^{\ell}(\alpha\beta;\ \tau < \ell)$$

$$= \mathbb{R}_{00}^{\ell}(\alpha;\ \tau < \ell;\ \mathbb{R}_{00}^{\ell}(\beta \mid F_\tau)).$$

By the strong Markov property

$$\mathbb{R}_{00}^{\ell}(\beta \mid F_\tau) = \mathbb{R}_{x0}^{\ell-\tau}(\beta \circ \theta_\tau),$$

and, consequently, we have

$$\mathbb{R}_{00}^{\ell}(\alpha\beta;\ M_\ell \geq x) = \int_0^\ell \mathbb{R}_{00}^{\ell}(\alpha;\ \tau \in ds)\ \mathbb{R}_{x0}^{\ell-s}(\beta \circ \theta_s)$$

$$= \int_0^\ell h(s,x)\ \mathbb{R}_0(\alpha;\ \tau \in ds)\ \hat{\mathbb{P}}_{x0}^{\ell-s}(\beta \circ \theta_s),$$

where h is given by (4.2) and $\hat{\mathbb{P}}_{x0}^{\ell}$ is the law of $\mathrm{ABMB}_{x0}^{\ell}$ (see (2.9)). We obtain

$$\mathbb{P}^e(\alpha\beta) = \int_0^\infty \mathbb{R}_0(\alpha;\ \tau \in ds) \int_s^\infty \hat{\mathbb{P}}_{x0}^{\ell-s}(\beta \circ \theta_s)\ b_x(\ell - s)\ d\ell,$$

where we have changed the order of integration and done some obvious reformulations. By (2.11) the latter integral equals $\hat{\mathbb{P}}_x(\beta \circ \theta_s)$, where $\hat{\mathbb{P}}_x$ is the law of ABM_x. Hence we have

$$\mathbb{P}^e(\alpha\beta) = \int_0^\infty \mathbb{R}_0(\alpha;\ \tau \in ds)\ \hat{\mathbb{P}}_x(\beta \circ \theta_s)$$

and the proof is complete.

Taking the time reversal of Proposition 1 we obtain

(2) COROLLARY. The process X^e is identical in law with the process Z_2 defined by

$$Z_2(t) = \begin{cases} R(t) & t \leqq \lambda_x, \\ x - \hat{R}(t) & \lambda_x > t, \end{cases}$$

where R is BES_0 and $\hat{R}$ is another BES_0 independent of R and killed when it reaches x.

Finally we arrive at Williams' characterization of the excursion law (see [15]).

(3) COROLLARY. The Brownian excursion process having a given maximum $x > 0$ is identical in law with the process Z_3 defined as

$$Z_3(t) = \begin{cases} R(t) & t \leqq \tau_x, \\ x - \hat{R}(t) & t \geqq \tau_x, \end{cases}$$

where R and $\hat{R}$ are as in Corollary 2.

PROOF. By Proposition 1 the beginning part of Z_3 is as it should be. Let $\varepsilon > 0$ and consider the Brownian part of the excursion given that the maximum of the excursion lies in $[x,x+\varepsilon)$. This conditioning converts ABM_x to a process which is identical in law with a process $x + \varepsilon - \hat{R}$, where $\hat{R}$ is BES_ε killed when it reaches $x + \varepsilon$. Let $\varepsilon \to 0$ to complete the proof.

6. Limit Theorems for Sojourn and Local Times

Let X be BM_0 and for a fixed $t > 0$ set

$$\sigma_t = \sup\{s<t: X(s) = 0\}$$

$$\eta_t = \inf\{s>t: X(s) = 0\}.$$

The process $X^e = \{X(\sigma_t+s): 0 \leq s \leq \eta_t - \sigma_t\}$ is called the excursion process straddling t. Let

$$L^t(x) = \lim_{\varepsilon \downarrow 0} \varepsilon^{-1} S^t((x,x+\varepsilon)),$$

where

$$S^t(A) = \int_{\sigma_t}^{\eta_t} 1_A(X_s)\, ds .$$

$L^t(x)$ is the ultimate value of the local time at x for X^e and $S^t(A)$ is the sojourn time for X^e in the set A.

In [2] the following two limit theorems are established:

1) The random variable $\dfrac{L^t(x)}{x}$ has a limiting distribution as $x \downarrow 0$. This distribution is the convolution of two exponential distributions each having mean 2, and, hence, independent of t.

2) The random variable $\dfrac{S^t((0,\varepsilon))}{\varepsilon^2}$ has a limiting distribution as $\varepsilon \downarrow 0$. This distribution is independent of t, and has the Laplace transform $(\sinh\sqrt{2\beta}\,)^{-2}$.

Theorem 2 goes back to Chung [1]. Proofs in [2] are rather long and tedious compared with the simplicity of the results. The advantage of the methods of Getoor and Sharpe are their generality. However, the point we want to make here is that Williams' characterization of the excursion law provides a proper tool to solve problems of this kind.

PROOF of (1). Let $y > 0$ be fixed and $\hat{X}$ BM_0 killed when it hits y. Denote the jointly continuous version of the local time of $\hat{X}$ with L. Then (see Knight [6] Theorem 1.3) $Y(x) = L_{\tau_y}(x) - L_{\lambda_0}(x)$, $0 \leq x \leq y$, is identical in law with the diffusion bridge from 0 to 0 and of length y constructed at the end of Section 1. On the other hand it is obvious that $Y(x)$ is just the local time of BES_0 killed when it reaches y.

Let $e = \{e(t), t \geq 0\}$ be a Brownian excursion with maximum y and $L^e(x)$ the ultimate value of its local time at x. Then by Williams' characterization

$$L^e(x) = L^1(x) + L^2(x), \tag{3}$$

where $L^1(x)$ and $L^2(x)$ are independent and identically distributed with $Y(x)$. By (1.12) $Y(x)$ is exponentially distributed with mean $\frac{2x(y-x)}{y}$. Consequently the limiting distribution as $x \downarrow 0$ of $\frac{Y(x)}{x}$ is exponential with mean 2, and, because this is independent of y, the representation (3) completes the proof.

(4) REMARK. Notice that we do not need the whole strength of Knight's theorem. In fact, it is quite clear that $L^i(x)$ must be exponentially distributed and its mean can be calculated from the definition of the local time. We leave this to the interested reader.

PROOF of (2). For the excursion e as above the total sojourn time splits in two independent parts:

$$S^e((0,\varepsilon)) = S^1((0,\varepsilon)) + S^2((0,\varepsilon)),$$

where S^1 is the total sojourn time in $(0,\varepsilon)$ of BES_0 killed when it hits y. By time reversal this is also the description for S^2; hence S^1 and S^2 are identically distributed.

Let R be $BES_{y-\varepsilon}$ killed when it hits y and $\hat{S}$ its total sojourn in $(y-\varepsilon,y)$. Then S^1 and S^2 are identically distributed with $\hat{S}$.

Next do a random time change of R with the additive functional

$$A_t = \int_0^t 1_{[y-\varepsilon,y)}(R_s)\, ds.$$

Denote the resulting process with $\hat{R}$. It is obvious that $\hat{S}$ is just the life time (i.e. the first hitting time of y) for the process $\hat{R}$.

Notice that $\hat{R}$ can also be described as $BES_{y-\varepsilon}$ having a reflecting barrier at $y - \varepsilon$ and killed when it hits y. Consequently,

$$E_{y-\varepsilon}(e^{-\beta\hat{S}}) = \mathbb{R}_{y-\varepsilon}(e^{-\beta\tau_y}) = \frac{\phi_1(y-\varepsilon)}{\phi_1(y)}, \quad \beta > 0,$$

where ϕ_1 is the increasing solution of

(5) $$\frac{1}{2}\frac{d^2u}{dx^2} + \frac{1}{x}\frac{du}{dx} = \beta u$$

with the conditions

(i) $$\frac{du}{dx}(y-\varepsilon) = 0$$

(ii) $$\lim_{x\uparrow\infty} u(x) = +\infty.$$

The general solution of (5) is

$$u(x) = \frac{1}{x}\,(c_1 e^{\sqrt{2\beta}\,x} + c_2 e^{-\sqrt{2\beta}\,x}).$$

After some calculations we obtain

$$\mathbb{R}_{y-\varepsilon}(e^{-\beta\tau_y}) = \frac{2\sqrt{2\beta}\;y}{(\sqrt{2\beta}(y-\varepsilon)+1)\;\exp(\varepsilon\sqrt{2\beta}) + (\sqrt{2\beta}(y-\varepsilon)-1)\;\exp(-\varepsilon\sqrt{2\beta})}$$

$$=: F(y,\varepsilon,\beta).$$

Consequently

$$\lim_{\varepsilon\downarrow 0} E_{y-\varepsilon}(e^{-\beta\hat{S}/\varepsilon^2}) = \lim_{\varepsilon\downarrow 0} \mathbb{R}_{y-\varepsilon}(e^{-\beta\,\tau_y/\varepsilon^2})$$

$$= \lim_{\varepsilon\downarrow 0} F(y,\varepsilon,\beta/\varepsilon^2)$$

$$= \frac{1}{\sinh\sqrt{2\beta}}\,.$$

Because S^1 and S^2 are independent and identically distributed and limit above is independent of y the proof of (2) is complete.

7. Lévy Measures

Let $X = \{X_t, \mathbb{P}_\bullet\}$ be a regular, canonically defined diffusion on $[0,+\infty)$. We assume that 0 is an entrance-exit point and give at 0 the boundary condition of instantaneous reflection. Further, we assume that the scale function s satisfies $s(0) = 0$ and $s(+\infty) = +\infty$. This implies that the local time of X at 0, denoted L_t, increases to the infinity as $t \to +\infty$.

We are interested in the right-continuous inverse of L, denoted by A. It is well known that A is an increasing Lévy process. Because of our boundary condition the law of A is completely determined by its Lévy measure n and this can be calculated from the formula

$$(1) \qquad n(d\ell) = \lim_{x \downarrow 0} \frac{\mathbb{P}_x(\tau_0 \in d\ell)}{s(x)} .$$

A proof of this can be found in [4] pp. 214-216. We present here a new proof based on the excursion theory and a generalization of the formula 4.5.d. We hope that our proof gives some intuitive insight to the reasons behind (1). The case studied by Itô and McKean is slightly more general than ours. However, it is not difficult to see that our proof applies with minor modifications also in their case.

To describe the excursion law of X consider the process $\hat{X} = (X_t, \hat{\mathbb{P}}_\bullet)$ obtained from X by killing it at the first hitting time of 0. Further, let $X^\uparrow = (X_t, \hat{\mathbb{P}}_\bullet)$ be the diffusion $\hat{X}$ conditioned never to hit 0. Then the following is true:

(2) a) 0 is an entrance, non-exit boundary point for $X^\uparrow$.

b) $$\mathbb{P}_x^\uparrow(f \circ X_t) = \frac{1}{s(x)} \hat{\mathbb{P}}_x((f \cdot s) \circ X_t), \qquad x > 0,$$

$$\mathbb{P}_0^\uparrow(f \circ X_t) = \lim_{x \downarrow 0} \mathbb{P}_x^\uparrow(f \circ X_t).$$

The excursion law ν of X can be described as follows (see [10](3.1))

(3) a) For $x > 0$ $\nu(m \geqq x) = 1/s(x)$

b) Given that the maximum of an excursion is greater than x to build up the excursion run $X^\uparrow$ up from 0 until it hits x. Then start an (independent) $\hat{X}$ diffusion.

Before proving (1) let us state a few facts from the theory of diffusions (see[4]). Let $\hat{X}$ and $X^\uparrow$ be as above. There exists two functions ϕ_1 and ϕ_2 such that

$$\text{(4)} \qquad \hat{E}_x(e^{-\alpha\tau_y}) = \begin{cases} \dfrac{\phi_1(x)}{\phi_1(y)}, & x \leqq y, \\[2ex] \dfrac{\phi_2(x)}{\phi_2(y)}, & x \geqq y. \end{cases}$$

Consequently, because 0 is a killing boundary for $\hat{X}$, $\phi_1(0) = \lim_{x\downarrow 0} \phi_1(x) = 0$. Further, by (2.b) above

$$\text{(5)} \qquad E_x^{\uparrow}(e^{-\alpha\tau_y}) = \begin{cases} \dfrac{\phi_1(x)}{\phi_1(y)} \cdot \dfrac{s(y)}{s(x)}, & x \leqq y, \\[2ex] \dfrac{\phi_2(x)}{\phi_2(y)} \cdot \dfrac{s(y)}{s(x)}, & x \geqq y. \end{cases}$$

Finally, ϕ_1 and ϕ_2 are positive, continuous, strictly increasing and decreasing, respectively, and solve the differential equation

$$\text{(6)} \qquad \frac{d}{dm}\frac{d}{ds}\,\phi = \alpha\phi.$$

PROOF of the formula (1). Let $x > 0$ be fixed and consider the finite measures

$$\mu_x^1(A) = \int_A \nu(m \geqq x \mid \zeta = \ell)\; \nu(\zeta \in d\ell)$$

$$\mu_x^2(A) = \nu(m \geqq x) \int_A \nu(\zeta \in d\ell \mid m \geqq x),$$

where $A \in \mathcal{B}\{[0,\infty)\}$. It is obvious that $\mu_x^1(I) = \mu_x^2(I)$ for all open intervals I. Consequently, by the monotone class theorem $\mu_x^1(f) = \mu_x^2(f)$ for all bounded, Borel-measurable functions f on $[0,\infty)$. Let $\alpha > 0$ and set $f(\ell) = 1 - e^{-\alpha\ell}$. Then we have

$$\int_0^\infty (1-e^{-\alpha\ell})\ \nu(m\geqq x \mid \zeta=\ell)\ \nu(\zeta\in d\ell) = \int_0^\infty (1-e^{-\alpha\ell})\ \nu(m\geqq x)\ \nu(\zeta\in d\ell \mid m\geqq x).$$

Let $x \downarrow 0$ and use the monotone convergence theorem to obtain

$$(7) \qquad \int_0^\infty (1-e^{-\alpha\ell})\ \nu(\zeta\in d\ell) = \lim_{x\downarrow 0} \int_0^\infty (1-e^{\alpha\ell})\ \nu(m\geqq x)\ \nu(\zeta\in d\ell \mid m\geqq x).$$

Now (3.a.b), the generalization of (4.5.d), and the fact $\nu(\zeta\in d\ell) = n(d\ell)$ allows us to rewrite (7) in the form

$$\int_0^\infty (1-e^{-\alpha\ell})\ n(d\ell) = \lim_{x\downarrow 0} \frac{1 - E_0^{\uparrow}(e^{-\alpha\tau_x})\ \hat{E}_x(e^{-\alpha\tau_0})}{s(x)} .$$

The claim is that

$$(8) \qquad \lim_{x\downarrow 0} \frac{1 - E_0^{\uparrow}(e^{-\alpha\tau_x})}{s(x)} = 0.$$

To prove this use (5) to obtain

$$\lim_{x\downarrow 0} \frac{1 - E_0^{\uparrow}(e^{-\alpha\tau_x})}{s(x)} = \lim_{x\downarrow 0} \frac{\phi_1(x) - \phi_x'(0)\ s(x)}{s(x)\ \phi_1(x)}$$

where $\phi_1'(0) = \lim_{x\downarrow 0} \frac{\phi_1(x)}{s(x)} > 0$. But we have

$$\phi_1(x) = \int_0^x \frac{d\phi_1}{ds}\ ds$$

$$= \int_0^x \left(\int_0^y \frac{d}{dm}\frac{d}{ds}\ \phi_1\ dm\right) ds + \phi_1'(0)\ s(x).$$

Use (6) and change the order of integration, the result is

$$\phi_1(x) - \phi_1'(0)\ s(x) = \int_0^x (s(x)-s(y))\ \alpha\phi_1(y)\ m(dy) > 0.$$

Therefore,

$$0 < \frac{\phi_1(x) - \phi_1'(0)\, s(x)}{(s(x))^2} \leq \int_0^x \alpha \frac{\phi_1(y)}{s(x)}\, m(dy)$$

$$\leq \int_0^x \alpha \frac{\phi_1(y)}{s(y)}\, m(dy) \to 0$$

as $x \downarrow 0$. Because $\phi_1'(0) > 0$ this gives (8). Hence

$$\int_0^\infty (1-e^{-\alpha\ell})\, n(d\ell) = \lim_{x\downarrow 0} \frac{1 - \hat{E}_x(e^{-\alpha\tau_0})}{s(x)}$$

Inverting Laplace transforms gives (1).

Acknowledgement

I wish to express my gratitude to Prof. E. Çinlar for his encouragement and support when preparing this paper. I am grateful to Prof. J.M. Harrison who provided me with (3.3).

References

1. K.L. CHUNG. Excursions in Brownian Motion. *Ark. Mat. 14,* 155-177 (1976).

2. R.K. GETOOR, M.J. SHARPE. Excursions of Brownian motion and Bessel processes. *Z. Wahrscheinlichkeitstheorie verw. Gebiete 47,* 83-106 (1979).

3. K. ITÔ. Poisson point processes attached to Markov processes. Proc *6th Berkeley Symp. Math. Statist. Prob., Vol. 3,* University of California Press, 225-240 (1971).

4. K. ITÔ, H. McKEAN. *Diffusion Processes and Their Sample Paths.* Springer-Verlag, Berlin, 1965.

5. N. IKEDA, S. WATANABE. *Stochastic Differential Equations and Diffusion Processes.* North-Holland, Amsterdam, 1981.

6. F.B. KNIGHT. Brownian local times and taboo processes. *Trans. Amer. Math. Soc. 143,* 173-185 (1969).

7. _____. Essentials of Brownian motion and diffusion. *Mathematical surveys 18.* Providence, Rhode Island: Amer. Math. Soc. (1981).

8. _____. On the excursion process of Brownian motion. *Trans. Amer. Math. Soc. 258,* 77-86 (1980).

9. H. McKEAN. Excursions of a non-singular diffusion. *Z. Wahrscheinlichkeitstheorie verw. Gebiete 1,* 230-239 (1963).

10. J. PITMAN, M. YOR. A decomposition of Bessel bridges. *Z. Wahrscheinlichkeitstheorie verw. Gebiete 59,* 425-457 (1982).

11. _____. Bessel Processes and infinitely divisible laws. *Stochastic Integrals,* Proc. LMS Durham Symposium, pp. 285-370, Lecture Notes in Math. 851. Springer-Verlag, Berlin, Heidelberg, New York, 1981.

12. L.C.G. ROGERS. Williams' characterization of the Brownian excursion law: proof and applications. *Séminaire de Probabilitiés XV,* pp. 227-250. Lecture notes in Math. 850. Springer-Verlag, Berlin, Heidelberg, New York, 1981.

13. P. SALMINEN. One-dimensional diffusions and their exit spaces. To appear in Math. Scand.

14. D. WILLIAMS. Decomposing the Brownian path. *Bull. Am. Math. Soc. 76,* (1970) 871-873.

15. _____. *Diffusions, Markov processes and martingales, Vol. 1: Foundations.* Wiley, New York, 1979.

P. SALMINEN
Åbo Academi
Matematiska Institutionen
SF-20500 ÅBO 50
FINLAND

Seminar on Stochastic Processes, 1983
Birkhäuser, Boston, 1984

CONSTRUCTION OF RIGHT PROCESSES FROM HITTING DISTRIBUTIONS

by

C. T. SHIH

0. Introduction

Let K be the one-point compactification of a locally compact second countable Hausdorff space and $\Delta \in K$ be the point at infinity. We are concerned with the problem of constructing Markov processes on K with Δ as the adjoined death point, from given hitting distributions. The most general Markov processes for the consideration of this problem (and indeed for the study of probabilistic potential theory) are those now known as right processes on a space K as above. (See [2].) It is well known that such a process is determined, up to a (random) time change, by its hitting distributions of compact sets of the state space. Our problem is therefore to construct a right process on K with prescribed hitting distributions $H_D(x,\cdot)$ for all compact $D \subset K$ and $x \in K$.

The problem, aside from being a fundamental one about the existence of Markov processes, arises naturally in connection with the axiomatic theory of harmonic functions (theory of harmonic spaces) developed by Brelot, Bauer and others. For reference to earlier papers on this problem by Meyer, Knight-Orey, Dawson, Boboc-Constantinescu-Cornea, and

Hansen see [4] and [5]. The processes constructed in these papers were mostly strong Feller processes. In [4] and later in [5] a more or less definitive treatment of this problem was given for the class of Feller processes.

This paper does away with the Feller conditions (continuity conditions that are nonintrinsic) of the measures $H_D(x,\cdot)$. We are able to construct any transient Hunt process (up to a time change) from its hitting distributions. This is done by first constructing a function e on K that determines the time scale; e(x) will be the expected lifetime of the procéss starting at x. The major part of the work is to construct a (unique) right process with given hitting distributions and a suitable function e. Here the quasi-left-continuity (which implies that the process is a Hunt process) is not assumed; thus we are able to construct all right processes with finite expected lifetime, requiring however the paths to have left limits. As in [4] and [5], only transient processes are constructed. To construct a recurrent process from given hitting distributions one can first obtain a sequence of locally defined (transient) processes from the results here, and then piece together these processes after making suitable time changes to form a globally defined process.

The work has retained much of the general approach in [4] and [5]; but all the crucial steps that have parallels in the earlier papers require new and often much harder proofs. (Aside from quoting one simple lemma in [4], the paper is completely independent of [4] and [5].) It is interesting to note that the method of compactification (in order to obtain a "Feller" family of hitting distributions on an enlarged state space so that earlier results become applicable), which might seem a natural approach, is not employed here. We suspect that such an approach may not be easier if it can be successful at all. Instead we use a direct attack which necessitates more probabilistic (and deeper) proofs.

Finally we mention the paper [3] by Gravereaux and Jacod, which to our knowledge is the only one published on this problem since [5]. It treats the restrictive class of processes with the space of reals as the state space and having paths continuous up to the lifetime; but it does use a quite different approach.

1. Main Results

As in the introduction K is a compact metric space, Δ is a fixed point in K and $\mathcal{D}$ denotes the family of closed sets in K containing Δ. Let d be the metric on K, $\mathcal{B}$ be the σ-algebra of Borel sets of K and $\mathcal{B}^*$ that of universally measurable sets of K (i.e. $\mathcal{B}^* = \cap\{\mathcal{B}^\mu : \mu$ a probability measure on $\mathcal{B}\}$ where $\mathcal{B}^\mu$ denotes the μ-completion of $\mathcal{B}$). Below we will introduce a σ-algebra $\mathcal{B}^n$ with $\mathcal{B} \subset \mathcal{B}^n \subset \mathcal{B}^*$; $b\mathcal{B}$, $b\mathcal{B}^n$ and $b\mathcal{B}^*$ stand respectively for the spaces of bounded real-valued $\mathcal{B}$-, $\mathcal{B}^n$- and $\mathcal{B}^*$-measurable functions on K. $\mathbf{C}(K)$ denotes the space of real-valued continuous functions on K.

A right process on K (with Δ as the adjoined heaven) is a strong Markov process $(X_t, \mathcal{F}_t, P^x)$ whose paths are right continuous and whose basic σ-algebras $\mathcal{F}_t$, relative to which the strong Markov property is defined and which are suitably completed from the minimal σ-algebras $\sigma(X_s, s \le t)$, are right continuous. See [1] and [2] for a complete definition and relevant facts. The right continuity of $\mathcal{F}_t$ (under the strong Markov property) is what we refer to as the intrinsic right continuity. The transition function $P(t,x,B)$ of a right process is only assumed to be universally measurable, i.e. $P(t,\cdot,B) \in b\mathcal{B}^*$ for $B \in \mathcal{B}$. We will deal with right processes whose paths have left limits; this is a reasonable restriction for our problem, especially because we treat only transient processes. For a right process $(X_t, \mathcal{F}_t, P^x)$ on K we define the (first) hitting time of a Borel set A to be $T_A = \inf\{t \geq 0: X_t \in A\}$; note the infimum is taken over nonnegative t rather than the usual strictly positive t. The hitting distribution of a Borel A for the process starting at x is the measure $P^x[X(T_A) \in \cdot, T_A < \infty]$.

Let $\{H_D(x,\cdot): D \in \mathcal{D}, x \in K\}$ be a family of measures on $\mathcal{B}$ (and therefore on $\mathcal{B}^*$). We introduce the following hypotheses on this family.

H1) $H_D(x,\cdot)$ *is a probability measure concentrated on* D *for all* x *and* D *(in* $\mathcal{D}$*), and is the point mass* ε_x *at* x *if* $x \in D$; $H_D(\cdot,B) \in b\mathcal{B}^*$ *for all* D *and for all* $B \in \mathcal{B}$ *(equivalently for all* $B \in \mathcal{B}^*$*).*

H2) *(Markov property). If* $D \subset D'$, $H_D(x,B) = \int H_{D'}(x,dy)H_D(y,B)$ *for all* x *and* $B \in \mathcal{B}^*$, *i.e.* $H_D f = H_{D'} H_D f$ *for all* $f \in b\mathcal{B}^*$ *where* $H_D f(y) \equiv \int H_D(y,dz)f(z)$.

Before stating the next hypothesis, we need to define the nearly Borel sets relative to the family $\{H_D(x,\cdot)\}$. A subset B of K is nearly Borel if for every probability measure μ on $\mathcal{B}$ there exist B_1, B_2 in $\mathcal{B}$ such that $B_1 \subset B \subset B_2$ and for all compact $C \subset B_2 - B_1$, $\int \mu(dx)\, H_{C \cup \Delta}(x,C) = 0$[1]. We will show in Theorem 2.11 that the family $\mathcal{B}^n$ of nearly Borel sets is a σ-algebra (using H1) and H2)); obviously $\mathcal{B} \subset \mathcal{B}^n \subset \mathcal{B}^*$.

H3) *(Nearly Borel measurability).* $H_D(\cdot,B) \in b\mathcal{B}^n$ *for all* D *and* $B \in \mathcal{B}$.

H4) *(Quasi-left-continuity). If* $D_n \downarrow D$ *(*D_n *decreases to* D*),* $H_{D_n}(x,\cdot)$ *converges weakly to* $H_D(x,\cdot)$, *i.e.* $H_{D_n} f(x) \to H_D f(x)$ *for all*

---

[1] Singletons $\{\Delta\}$ and $\{x\}$ are often written as Δ and x for convenience.

$f \in \mathbf{C}(K)$, *for any* x.

H4A) *For any* x *and* $D_n \downarrow D$ *the following are satisfied:*

H4A.1) $H_{D_n}(x,\cdot)$ *converges weakly;*

H4A.2) *if compact sets* $F_m \uparrow K - D$ *then for* $\varepsilon > 0$ *there is* $\delta > 0$ *such that for all* m *for which the weak limit*

$$\nu_m(dy,dz) = w - \lim_n H_{D_n}(x,dy) H_{F_m \cup D}(y,dz)$$

exists, $\nu_m\{(y,z): 0 < d(y,z) < \delta\} < \varepsilon$.

H5) *(Intrinsic right continuity)*[1]. *Fix* x *and an increasing sequence* D_n, *and let* (W_n) *be the nonhomogeneous reversed Markov chain (under a single probability measure* P*) on* $(K, \mathcal{B}^*)$ *satisfying* $P(W_n \in \cdot) = H_{D_n}(x,\cdot)$ *and* $P(W_n \in \cdot \mid W_m, m > n) = H_{D_n}(W_{n+1},\cdot)$. *If* $W_\infty = \lim_n W_n$ *exists a.s., then for any* $F \in \mathbf{D}$ *and* $f \in b\mathcal{B}^*$, $H_F f(W_n)$ *converges to* $H_F f(W_\infty)$ *a.s. on* $\{W_\infty \notin F\}$; *in particular, with* $\mu(\cdot) = P(W_\infty \in \cdot\,) = w - \lim_n H_{D_n}(x,\cdot)$, $H_{D_n}(x,dy)H_F f(y)$ *converges vaguely to* $\mu(dy)H_F f(y)$ *on the locally compact* $K - F$.

H6) *(Transience). For any* x *and* D *with* $x \notin D$, *there exists* D' *containing* x *in its interior such that* $\int H_D(x,dy) H_{D'}(y, D' - \Delta) < 1$.

THEOREM 1. *Let* $\{H_D(x,\cdot): D \in \mathbf{D},\ x \in K\}$ *be a family of measures on* $\mathcal{B}$ *satisfying* H1), H2), H3), H4), H5) *and* H6). *Then there exists a right process* $(H_t, \mathcal{F}_t, P^x)$ *on* K, *with* Δ *as an absorbing point, such that starting at any* x *its hitting distribution of any* $D \in \mathbf{D}$ *is* $H_D(x,\cdot)$.

---

[1] It is possible to state H5) in purely "analytic" terms; but this version is more direct. See the footnote in Theorem 2.10 about the existence of (W_n).

The process in Theorem 1 is not unique (up to equivalence) since the time scale is not prescribed. The next theorem deals with the construction of right processes having a prescribed time scale as well as prescribed hitting distributions. Under the conditions of Theorem 1 we can define a (nonunique) function e on K satisfying the hypothesis below, with $e(x)$ meant to be the expected lifetime (time of reaching Δ) of the process starting at x. For such a function e we define

$$e_D(x) = e(x) - H_D e(x) = e(x) - \int H_D(x,dy)e(y).$$

Of course, with the intended meaning of $H_D(x,\cdot)$ and $e(y)$, $e_D(x)$ stands for the expected hitting time of D.

Let e be a real-valued function on K; we introduce hypothesis:

H7) e *is a nonnegative* $\mathcal{B}^*$*-measurable function with* $e(\Delta) = 0$ *and satisfies the following conditions:*

H7.1) *For any* x, *neighborhood* U *of* x *and an increasing sequence* D_n, *if* $\inf_n H_{D_n}(x, K-U) > 0$ *then* $\inf_n e_{D_n}(x) > 0$.

H7.2) *If* x *is instantaneous (by definition* $x \neq \Delta$ *and satisfies the property that if* $D_n \uparrow K - \{x\}$ *with the closure* $\overline{K - D_n} \downarrow \{x\}$, $H_{D_n}(x,\cdot)$ *converges weakly to the point mass* ε_x), $\inf\{e_D(x) : x \notin D\} = 0$.

H7.3) *If* $D_n \downarrow D$ *and compact sets* $F_m \uparrow K - D$, *then for any* x

$$\lim_m \overline{\lim_n} \int H_{D_n}(x,dy)\, e_{F_m \cup D}(y) = 0.$$

H7.4) *For any* x *and* $\varepsilon > 0$ *there exists* $a > 0$ *such that* $H_D e(x) < \varepsilon$ *for all* D *with* $D - \Delta \subset \{y : e(y) \geq a\}$.

THEOREM 2. *Let* $\{H_D(x,\cdot) : D \in \mathcal{D},\ x \in K\}$ *be a family of measures on* $\mathcal{B}$ *satisfying* H1, H2), H3), H4A), *and* H5), *and* e *be a real-valued function on* K *satisfying* H7). *Then there exists a unique right process on* K, *with* Δ *as the death point, such that starting at any* x *its hitting distribution of any* $D \in \mathcal{D}$ *is* $H_D(x,\cdot)$ *and its expected lifetime equals* e(x).

The following discussions about the various hypotheses should serve to clarify them and help understanding the above theorems. First, on hypothesis H1): That all the measures $H_D(x,\cdot)$ have total mass 1 forces the process to be constructed to have finite lifetime, since $\{\Delta\} \in \mathcal{D}$. If one is given subprobability measures $H_D(x,\cdot)$, they can always be made probability measures by adding mass at Δ, so as to make the results applicable. That $H_D(x,\cdot) = \varepsilon_x(\cdot)$ if $x \in D$ follows from the way we defined hitting times. The universal measurability in H1) is a preliminary requirement; without it we cannot state H2) and define the σ-algebra $\mathcal{B}^n$. The reader willing to accept the Borel measurability of the kernels $H_D(x,B)$ $(H_D(\cdot,B) \in b\mathcal{B}$ if $B \in \mathcal{B})$ may simply read $\mathcal{B}^*$ (and of course $\mathcal{B}^n$) as $\mathcal{B}$ throughout the paper.

On H2): This is a consequence of the strong Markov property. However, the full strength of the strong Markov property (of a right process) is carried by H2) and H5) together. Note that in H2) B and f need only be required to belong to $\mathcal{B}$ and $b\mathcal{B}$ respectively.

On H3): H3) is necessary in a right process. We show in Theorem 2.11 that H3) is equivalent, of course under H1) and H2), to: $H_D(\cdot,B) \in b\mathcal{B}^n$ for all D and $B \in \mathcal{B}^n$. This hypothesis is used in a technical way in Lemma 2.8 (and three similar places: Theorems 2.9, 2.10 and 3.4) where a capacity argument is used. It is interesting to note that under H6B) below Lemma 2.8 and Theorems 2.9, 2.10 and 3.4 can be proved (relatively easily) without capacity arguments and therefore without H3).

On H4) and H4A): It will be proved in Theorem 2.9 that under H1), H2), H3) and H6), H4) implies condition H4A.2) and therefore hypothesis H4A). Of course condition H4A.1) reflects the requirement that paths have left limits. Let us now observe the necessity of H4A.2) in a right process. Clearly $\nu_m(dy,dz) = P^x(X(T-) \in dy,\ X(T_m)\circ\theta_T \in dz)$ where $T = \lim_n T_{D_n}$, $T_m = T_{F_m \cup D}$ (both are finite a.s.) and θ_t denotes the shift operator. Since for all sufficiently large m, $T_m \circ \theta_T = 0$ so that $d(X(T-), X(T_m)\circ\theta_T)$ is constant, H4A.2) must hold. The measures ν_m in H4A.2) actually always exist; see a remark following Theorem 2.9. In §8 we collect several pathological examples to illustrate the independence among the various conditions; Example 1 provides a family $\{H_D(x,\cdot)\}$ and a function e satisfying all conditions of Theorem 2 except H4A.2). Hypothesis H4) only plays a role in constructing a function e satisfying H7); specifically it is used in proving conditions H7.1) and H7.3), with its role in establishing H7.3) a crucial one. It is an open question whether a function e satisfying H7) can be defined with H4) replaced by the necessary H4A).

On H5): As in H2), we need only require f to belong to $b\mathcal{B}$. This hypothesis is an easy consequence of the intrinsic right continuity of a right process mentioned earlier. Indeed, the hitting times $T_n = T_{D_n} \downarrow T$; in particular $X_{T_n} \to X_T$. Using the right continuity $\mathcal{F}(T_n) \downarrow \mathcal{F}(T)$, the strong Markov property and a martingale argument one can show $H_F f(X_{T_n}) \to H_F f(X_T)$ a.s. P^x on $\{X(T) \notin F\}$, which is just the conclusion in H5). Of course, for the reader familiar with the fundamentals of Markov processes, H5) simply follows from the fact that $H_F f$ for $f \geq 0$ is excessive on $K - F$ and therefore a.s. $H_F f(X_t)$ is right continuous on intervals of t with $X_t \in K - F$. In H5) the a.s. limits of W_n and $H_F f(W_n)$ (if $W_\infty \notin F$) always exist; see Theorem 2.10. H5) is used in two places: one in proving condition H7.1) for the function e defined in §2; the other in proving the (strong) Markov property of the process constructed for Theorem 2. A special case of H5) is

H5A) *(Intrinsic right continuity at 0, or Blumenthal zero-one law). Fix* x *and a sequence* D_n *with* $D_n \uparrow K - \{x\}$ *and* $\overline{K - D_n} \downarrow \{x\}$, *and let* (W_n) *be defined from* D_n *as in* H5). *The statement in* H5) *about* (W_n) *holds.*

This condition implies the following fact (proved in Corollary 2.4): if D_n and x are as above, then $\mu(\cdot) = w - \lim_n H_{D_n}(x,\cdot)$ (the convergence of W_n in this case is immediate from H2)) satisfies $\mu\{x\} = 1$ or 0. As has been defined in the statement of condition H7.2), $x \neq \Delta$ is instantaneous if the weak limit μ of $H_D(x,\cdot)$ as $D \uparrow K - \{x\}$ with $\overline{K - D} \downarrow \{x\}$ satisfies $\mu\{x\} = 1$; thus $x \neq \Delta$ is noninstantaneous if this measure μ satisfies $\mu\{x\} = 0$. Examples 2 and 3 of §8 give families $\{H_D(x,\cdot)\}$ satisfying all conditions of Theorem 1 except H5); the one in Example 2 (which is trivial) does not satisfy H5A) but the other does.

On H6): Under H4) (and H1), H2)), H6) is equivalent to the otherwise weaker

H6A) *If* $x \notin D$, $\int H_D(x,dy)\, H_{x \cup \Delta}(y,x) < 1$.

This is because $H_{D'}(y, D' - \Delta)$ is increasing in D', an easy consequence of H2). H6A) may be called non-point-recurrence while H6) non-neighborhood-recurrence. Under H7) it is easily seen (from H7.1A) below) that H6A) is valid but H6) not necessarily. Note that H6) is not assumed in Theorem 2. Consider on the other hand the following stronger non-recurrence condition than H6):

H6B) *If* $x \notin D$, *there exist* D' *containing* x *in its interior, a neighborhood* U *of* x *and* $\varepsilon > 0$ *such that for all* $y \in U$,

$$\int H_D(y,dz)H_{D'}(z, D' - \Delta) < 1 - \varepsilon.$$

If we assume H6B) in either Theorem 1 (replacing H6)) or Theorem 2, then as mentioned earlier H3) does not have to be assumed; also condition H4A.1) need not be assumed in Theorem 2. (We will not elaborate on this.)

On H7): First, in a right process $e(x) \equiv P^x(T_\Delta)$ is nearly Borel measurable; but it is not necessary to make this stronger (but necessary) measurability assumption. However the function $e(x)$ defined for Theorem 1 is indeed $\mathcal{B}^n$-measurable as a consequence of H3). Condition H7.1) is necessary in a right process since $T_{K-U} > 0$ a.s. P^x and $T_D \geq T_{K-U}$ on $\{X(T_D) \in K - U\}$. Of course it is understood the existence of branching points is precluded;[1)] indeed H7.1) fails to hold at such points x. H7.1) implies the following condition (by taking $U = K - D$ and $D_n \equiv D$)

H7.1A) $\qquad e_D(x) > 0 \quad \text{if} \quad x \notin D.$

(We are thus entitled to use H7.1A) in proving Theorem 2.) Example 4 of of §8 provides a family $\{H_D(x,\cdot)\}$ and a function e satisfying H7.1A) and all conditions of Theorem 2 except H7.1) at an instantaneous point x. Let us now observe that under H7.1), $h(x) \equiv \inf\{e_D(x): x \notin D\}$ is positive if $x \neq \Delta$ is non-instantaneous (of course $h(x)$ stands for the expected holding time at such a point x). First, e_D decreases as D increases; for if $D \subset D'$, $e_{D'} = e - H_{D'}e = e - H_{D'}(e_D + H_De) = e - H_{D'}e_D - H_De = e_D - H_{D'}e_D \leq e_D$ since $e_D \geq 0$ ($e_D = 0$ on D and $e_D > 0$ on $K - D$ by H7.1A)). Thus $h(x)$ is the decreasing limit of

[1)]Theorem 2 holds with no changes of proofs if we do permit the existence of branching points, which will be the noninstantaneous points failing to satisfy H7.1). H7.1) will then be required for instantaneous points only. Of course one must also hypothesize that all measures $H_D(x,\cdot)$ do not charge the set of branching points.

$e_D(x)$ as $D \uparrow K - \{x\}$ (whether x is noninstantaneous or not). Now if $x \neq \Delta$ is noninstantaneous, then for some neighborhood U of x the weak limit μ of $H_D(x,\cdot)$ as $D \uparrow K - \{x\}$ with $\overline{K-D} \downarrow \{x\}$ satisfies $\mu(K-U) > 0$; it is thus clear that $h(x) > 0$ under H7.1).

Condition H7.2) is trivially necessary (the infimum is the $h(x)$ defined above), and the equally trivial Example 5 of §8 (given for the sake of completeness) has a family $\{H_D(x,\cdot)\}$ and a function e satisfying all conditions of Theorem 2 except H7.2).

About condition H7.3), the limit is $\lim_m P^x(T_{F_m \cup D} \circ \theta_T)$ [1] where $T = \lim_n T_{D_n}$; since $T_{F_m \cup D} \downarrow 0$ a.s., this limit must be 0. The trivial Example 6 in §8 gives a family $\{H_D(x,\cdot)\}$ and a function e satisfying all conditions of Theorem 2 except H7.3). The function e defined for Theorem 1 satisfies

H7.3A) If $D_n \downarrow D$, $e_{D_n}(x) \uparrow e_D(x)$ for all x.

This condition implies H7.3) since $\int H_{D_n}(x,dy)e_{F_m \cup D}(y) \leq \int H_{D_n}(x,dy)e_D(y)$ which one checks to be $e_{D_n}(x) - e_D(x)$. It is not difficult to show that a right process (with a finite expected lifetime function e) satisfying H7.3A) must satisfy H4).

About condition H7.4), first note that a bounded e certainly satisfies this condition, and such is the case for the function e defined for Theorem 1. Let us explain the necessity of H7.4) in a right process with a finite expected lifetime function e. If it fails, then for some x and $\varepsilon > 0$ we have for all $n > 0$, $H_{D_n}e(x) = P^x(T_\Delta - T_{D_n}) \geq \varepsilon$ for some D_n with $D_n - \Delta \subset \{y\colon e(y) \geq n\}$. Let $A_n = \bigcup_{k \geq n} D_k$; then $P^x(T_\Delta - T_{A_n}) \geq \varepsilon$ for all n. But clearly $P^x(T_{A_n} < T_\Delta) = \lim_k P^x(T_{D_n \cup \cdots \cup D_k} < T_\Delta) \leq e(x)/n$; by the dominated convergence we

---

[1] Throughout the paper, expectations such as $\int \phi dP^x$ are denoted by $P^x(\phi)$ and $\int_\Lambda \phi dP^x$ by $P^x(\phi;\Lambda)$.

then have the contradiction $P^x(T_\Delta - T_{A_n}) \not\downarrow 0$. Example 7 in §8 provides a family $\{H_D(x,\cdot)\}$ and a function e satisfying all conditions of Theorem 2 except H7.4).

The steps of proofs of Theorems 1 and 2 are as follows. In §2 it is proved that conditions of Theorem 1 imply those of Theorem 2. In §§3-7 we prove Theorem 2. This is done by first constructing a process Z_∞ that contains all information of the trajectories of the sought-for process (X_t). Z_∞ will be the projective limit of a sequence of discrete parameter Markov processes (chains) $Z_n = (Z_{n\alpha})$ where the time parameter α takes on ordinal values less than a certain countable ordinal. A typical Z_n is studied in §3; in §4 we define Z_∞ and examine some of its properties. In §5 we define functions $S(n,\alpha)$[1)] in Z_∞ which determine the time scale of the sought-for process; a crucial property that $S(n,\alpha)$ is strictly increasing in α is established in §6. Finally in §7 we define the process (X_t) (under the assumption no holding points exist while in the general case the (X_t) defined has holding times in integrated form and we omit the last purely technical step of defining the legitimate desired right process), show that it has the correct hitting distributions and prove the strong Markov property. In §§5-7 proofs in the case when holding points are present are given somewhat sketchily.

2. The function e

In this section we consider a family $\{H_D(x,\cdot): D \in \mathcal{D}, x \in K\}$ of measures on K satisfying conditions of Theorem 1 (except for Theorems 2.10 and 2.11). A function e will be defined on K and shown to

---

[1)] In [4] functions similar to $S(n,\alpha)$ were defined as the limits of certain martingales; but the martingale assertion there was false. Here the mistake is corrected.

satisfy H7). Also, hypothesis H4A) will be shown to hold. Therefore Theorem 1 will follow from Theorem 2.

Let $\mathbf{E}$ consist of all mappings $r \to D(r)$, $0 \le r \le 1$, such that $D(r) \in \mathbf{D}$ and $D(r) - \Delta$ is compact for each r, and $D(r_1) - \Delta \subset (D(r_2) - \Delta)^\circ$ if $r_1 < r_2$ (A° denotes the interior of A). Choose a sequence of mappings $F_k(r)$ in $\mathbf{E}$ such that for all $x \ne \Delta$ and $\delta > 0$ there is some k with $x \in F_k(0) - \Delta \subset F_k(1) - \Delta \subset B(x,\delta)$, where $B(x,\delta) = \{y\colon d(x,y) < \delta\}$. Define e as follows:

$$e(x) = \sum_k 2^{-k} \int_0^1 H_{F_k(r)}(x, F_k(r) - \Delta)\,dr.$$

Since $H_D(x, D - \Delta)$ is increasing in D (if $D \subset D'$, $H_D(x, D - \Delta) = \int H_{D'}(x,dy) H_D(y, D - \Delta) \le H_{D'}(x, D' - \Delta)$), $H_{F_k(r)}(x, F_k(r) - \Delta)$ is increasing in r. Thus e is well-defined and in $b\mathbf{B}^*$ (in $b\mathbf{B}$ if the kernels $H_D(\cdot,\cdot)$ are $\mathbf{B}$-measurable). Actually by H3) $e \in b\mathbf{B}^n$; but this is not required in H7). Of course $e \ge 0$ and $e(\Delta) = 0$.

THEOREM 2.1. *The function* e *satisfies condition* H7.1).

PROOF. First, from Lemma 2.1 of [4], for arbitrary D, F

$$H_F(x, F - \Delta) \ge \int H_D(x,dy)\, H_F(y, F - \Delta). \tag{2.1}$$

Now

$$\begin{aligned} e_D(x) &= e(x) - \int H_D(x,dy) e(y) \\ &= \sum_k 2^{-k} \int_0^1 [H_{F_k(r)}(x, F_k(r) - \Delta) - \int H_D(x,dy) H_{F_k(r)}(y, F_k(r) - \Delta)]\,dr. \end{aligned}$$

By (2.1) the integrands in the above sum are nonnegative. In particular $e_D \ge 0$ and consequently as observed in the discussion about H7.2) e_D decreases as D increases. Now if D_n increases and $e_{D_n}(x) \downarrow 0$, then for all k

$$(2.2) \qquad H_{F_k(r)}(x,F_k(r)-\Delta) - \int H_{D_n}(x,dy)\, H_{F_k(r)}(y,F_k(r)-\Delta) \downarrow 0$$

a.e. dr as $n \to \infty$ (that the above is decreasing also follows from (2.1)). To prove H7.1), suppose it fails, so that for some x, neighborhood U of x and an increasing sequence D_n we have $\inf_n H_{D_n}(x,K-U) = \varepsilon > 0$ but $e_{D_n}(x) \downarrow 0$. Obviously we may assume $\Delta \in K-U$ and replace D_n by $D_n \cup (K-U)$ (because of H2)), and thus may assume $D_1 = K-U$; note then $H_{D_n}(x,D_1) \downarrow \varepsilon$ (for $B \subset D_1$, $H_{D_n}(x,B)$ decreases by H2)). Let D' contain x in its interior; since $e_{D_n}(x) \downarrow 0$ we can apply (2.2) to a mapping $F_k(r)$ with $x \in F_k(0)-\Delta \subset F_k(1)-\Delta \subset D'-\Delta$ to obtain (noting $H_{F_k(r)}(x,F_k(r)-\Delta) = 1$)

$$(2.3) \qquad \int H_{D_n}(x,dy)\, H_{D'}(y,D'-\Delta) \uparrow 1.$$

From (2.3) we will show that

$$\int H_{D_1}(x,dz)\, H_{D'}(z,D'-\Delta) = 1.$$

Since D' is arbitrary, this is a contradiction to H6). Now to prove this equality it suffices to show

$$(2.4) \qquad \int H_{D_n}(x,dy) \int H_{D_1}(y,dz)\, H_{D'}(z,D'-\Delta) \to 1.$$

We will establish separately in Theorem 2.10 that the (W_n) in H5) defined from D_n converges a.s. so that $\mu(\cdot) = w - \lim_n H_{D_n}(x,\cdot)$ exists, and in Lemma 2.3 that $\mu(D_1) = \lim_n H_{D_n}(x,D_1)$. Since $H_{D_n}(x,B)$ decreases for $B \subset D_1$, we must have $H_{D_n}(x,B) \downarrow \mu(B)$ for such B ($\mu_1(B) = \lim_n H_{D_n}(x,B)$ defines a measure on D_1 which must agree with μ there). Therefore $\int_{D_1} H_{D_n}(x,dy)f(y) \to \int_{D_1} \mu(dy)f(y)$ for $f \in b\mathcal{B}^*$. (Incidentally $\mu\{\Delta\} = 0$ by this and (2.3).) Combining this and H5)[1)] with $F = D_1$ and

[1)] Here and below in this section when H5) is used, we need only the vague convergence of $H_{D_n}(x,dy)H_F f(y)$ to $\mu(dy)H_F f(y)$ as measures on $K-F$, not the full force of H5).

$f(z) = H_{D'}(z, D' - \Delta)$ (noting $f = H_{D_1} f$ on D_1), we see that (2.4) will follow from

$$\int \mu(dy) \int H_{D_1}(y,dz)\, H_{D'}(z, D' - \Delta) = 1.$$

To establish this last equality for any D' containing x in its interior it suffices to prove

LEMMA 2.2. *The measure* μ *defined above satisfies*

$$\int \mu(dy) \int H_{D_1}(y,dz)\, H_{x \cup \Delta}(z,x) = 1.$$

PROOF. Let D' contain x in its interior and satisfy $\mu(\partial D') = 0$ (note $\mu\{\Delta\} = 0$ as observed above). (∂A denotes the boundary of A.) Using H5) with $F = D'$ and $f = 1_{D' - \Delta}$ and noting $H_{D'}(\cdot, D' - \Delta) = 1$ on $D' - \Delta$, we obtain from (2.3) that $\int \mu(dy)\, H_{D'}(y, D' - \Delta) = 1$. Let D' go through a sequence with $D' - \Delta$ compact and decreasing to x; we then have $H_{x \cup \Delta}(y,x) = 1$ a.e. $\mu(dy)$. It follows that

$$\int \mu(dy) \int H_{D_1 \cup x}(y,dz)\, H_{x \cup \Delta}(z,x) = 1. \tag{2.5}$$

Now $\int \mu(dy)\, H_{D_1 \cup x}(y, D_1) \geq \mu(D_1) = \lim_n H_{D_n}(x, D_1) = \varepsilon$. Since

$$\int \mu(dy) H_{D_1}(y,dz) = \int \mu(dy) H_{D_1 \cup x}(y,dz) 1_{D_1}(z) + \int \mu(dy) H_{D_1 \cup x}(y,x) H_{D_1}(x,dz)$$

and since the fact (which follows from observations above) that

$$H_{D_1}(x,dz) = \int H_{D_n}(x,dy) H_{D_1}(y,dz) \to \int \mu(dy) H_{D_1}(y,dz)$$

strongly (weakly is sufficient) implies $H_{D_1}(x,dz) = \int \mu(dy) H_{D_1}(y,dz)$, we have

$$\int \mu(dy) H_{D_1}(y,dz) = \int \mu(dy) H_{D_1 \cup x}(y,dz) 1_{D_1}(z) / \int \mu(dy) H_{D_1 \cup x}(y, D_1).$$

The lemma now follows from (2.5). □

The following lemma was used in the proof of Theorem 2.1.

LEMMA 2.3. *Let* x *be fixed and* D_n *increase; assume the* W_n *defined in* H5) *converges a.s. and* $\mu(\cdot) = w - \lim_n H_{D_n}(x,\cdot)$. *Then*

(i) $\mu(D_1) = \lim_n H_{D_n}(x,D_1)$;

(ii) *(without assuming* H4) *and* H6)) *if* $\lim_n H_{D_n}(x,D_1) > 0$ *then* $\mu\{x\} = 0$.

PROOF. To prove (i) it suffices to show $\lim_n H_{D_n}(x,D_1) \geq \mu(D_1)$. Let compact sets $C_k \downarrow D_1$ with $D_1 \subset C_k^o$ and $H_{D_n}(x,\partial C_k) = 0$ for all n, k. Using H2) it is easy to check that for $n < N$ and any k

$$H_{C_k \cup D_n}(x,C_k) \geq H_{C_k \cup D_N}(x,C_k) \geq H_{D_N}(x,C_k).$$

Since $\underline{\lim}_N H_{D_N}(x,C_k) \geq \mu(D_1)$ we have $H_{C_k \cup D_n}(x,C_k) \geq \mu(D_1)$ for all n, k. From H4) and the fact $H_{D_n}(x,\partial C_j) = 0$ it then follows that

$$H_{D_n}(x,D_1) = \lim_j H_{D_n}(x,C_j) = \lim_j \lim_k H_{C_k \cup D_n}(x,C_j)$$
$$\geq \overline{\lim_k} H_{C_k \cup D_n}(x,C_k) \geq \mu(D_1).$$

To prove (ii) let $\varepsilon = \lim_n H_{D_n}(x,D_1) > 0$. Apply H5) to $F = D_m$, $f = 1_{D_1}$ we have

$$H_{D_n}(x,D_1) = \lim_n \int H_{D_n}(x,dy)\, H_{D_m}(y,D_1)$$
$$\geq \lim_n [H_{D_n}(x,D_1) + \int_{K-D_m} H_{D_n}(x,dy)\, H_{D_m}(y,D_1)]$$
$$\geq \varepsilon + \int_{K-D_m} \mu(dy)\, H_{D_m}(y,D_1) \geq \varepsilon + \mu\{x\}\cdot H_{D_m}(x,D_1).$$

As $m \to \infty$ we have $\varepsilon \geq \varepsilon(1 + \mu\{x\})$; so (ii) follows. □

The result in the following corollary was mentioned in §1 when discussing the definition of instantaneous and noninstantaneous points.

COROLLARY 2.4. *Assume only* H1), H2), H3) *and* H5A). *If* $D_n \uparrow K-\{x\}$ *with* $\overline{K-D_n} \downarrow \{x\}$, *then* $\mu = w - \lim_n H_{D_n}(x,\cdot)$ *exists and* $\mu\{x\} = 1$ *or* 0.

PROOF. Let (W_n) be defined from D_n as in H5). From H2) it is clear that a.s. either $W_n \to x$ or W_n is constant for all large n. So μ exists. If $\mu\{x\} < 1$, then $\lim_n H_{D_n}(x,D_m) > 0$ for some m; by (ii) of the above lemma (whose proof relies only on H5A) for the D_n here) we then have $\mu\{x\} = 0$. □

THEOREM 2.5. *The function* e *satisfies condition* H7.2).

PROOF. Let x be instantaneous. By the monotonicity of $e_D(x)$ in D it suffices to prove $e_{D_n}(x) \downarrow 0$ for a sequence $D_n \uparrow K-\{x\}$ with $\overline{K-D_n} \downarrow \{x\}$. We must show

$$\int H_{D_n}(x,dy)\, H_{F_k(r)}(y,F_k(r)-\Delta) \to H_{F_k(r)}(x,F_k(r)-\Delta)$$

a.e. dr for all k (see (2.2)). Since for each k there is at most one value of r for which $x \in \partial F_k(r)$, it suffices to prove the above convergence when $x \notin \partial F_k(r)$. If $x \notin F_k(r)$ then for n sufficiently large $F_k(r) \subset D_n$ so that the two sides of the above display are equal. If $x \in F_k(r)^{\circ}$ then since $H_{D_n}(x,\cdot)$ converges to the point mass at x, $H_{D_n}(x,F_k(r)-\Delta) \to 1$ and the above convergence follows. □

THEOREM 2.6. *The function* e *satisfies the condition that if* $D_n \downarrow D$, $e_{D_n} \uparrow e_D$; *therefore it satisfies condition* H7.3) *(see the discussion about* H7.3A) *in* §1).

PROOF. Let $D_n \downarrow D$ and x be fixed. We show $\int H_{D_n}(x,dy)e_D(y) \to 0$, which is equivalent to $e_{D_n}(x) \uparrow e_D(x)$. Since for each k $H_D(x,\partial(F_k(r)-\Delta)) = 0$ except possibly for countably many r, it suffices to prove (see the beginning of the proof of Theorem 2.1)

$$(2.6) \qquad \int H_{D_n}(x,dy)[H_F(y,F-\Delta) - \int H_D(y,dz)H_F(z,F-\Delta)] \to 0$$

for arbitrary $F \in \mathcal{D}$ with $F-\Delta$ compact and $H_D(x,\partial(F-\Delta)) = 0$. Part of the proof will be contained in the next two lemmas. For the first lemma define a nonhomogeneous Markov chain $(Z_n, n \geq 0)$ on $(K,\mathcal{B}^*)$ (with a single probability measure P) satisfying $P(Z_0 = x) = 1$ and $P(Z_n \in B | Z_0,\ldots,Z_{n-1}) = H_{D_n}(Z_{n-1},B)$ for $n \geq 1$. From H2) we have $P(Z_n \in B | Z_0,\ldots,Z_k) = H_{D_n}(Z_k,B)$ for $k < n$; in particular $P(Z_n \in \cdot) = H_{D_n}(x,\cdot)$.

LEMMA 2.7. (i) *(Assuming condition* H4A.1) *in place of* H4)*),* Z_n *converges a.s.* dP. (ii) *For any* $\delta > 0$, $\int H_{D_n}(x,dy)H_D(y,B(y,\delta)) \to 1$ *(recall* $B(y,\delta) = \{z: d(y,z) < \delta\}$).

PROOF. Suppose (i) fails; then by compactness there exist z_0, $\delta > 0$ and a sequence $n_k \to \infty$ such that for each k

$$(2.7) \quad P(Z_{n_k} \in B;\ d(Z_{n_k},Z_n) > 2\delta \text{ for some } n,\ n_k < n < n_{k+1}) > \delta$$

where $B = B(z_0,\delta)$. Let $\mu = w - \lim_n H_{D_n}(x,\cdot)$ and assume as we may $\mu(\partial B) = 0$. Let $C_{2k-1} = D_{n_k}$ and $C_{2k} = (C_{2k-1} - B) \cup C_{2k+1}$. Then $H_{C_k}(x,\cdot)$ converges weakly to μ and so $\lim_k H_{C_k}(x,K-B)$ exists. But

$$H_{C_{2k}}(x,K-B) = H_{C_{2k-1}}(x,K-B) + \int_B H_{C_{2k-1}}(x,dy)H_{C_{2k}}(y,K-B).$$

Using H2) and an induction argument one can show that the second term on the right is at least $P(Z_{n_k} \in B;\ Z_n \notin B$ for some $n;\ n_k < n \leq n_{k+1})$, which by (2.7) is greater than δ. Thus we have the contradiction that

$H_{C_k}(x,K-B)$ does not converge, proving (i). From (i) and H4)

$$P(\lim_m Z_m \in dz|Z_n) = w - \lim_m H_{D_m}(Z_n,dz) = H_D(Z_n,dz).$$

Since $Z_n \to \lim_m Z_m$ a.s. and $P(Z_n \in dy) = H_{D_n}(x,dy)$, (ii) follows. □

To continue the proof of Theorem 2.6, define another nonhomogeneous chain $(Y_m,m \geq 0)$ by requiring $P(Y_0 = x) = 1$, $P(Y_{2n-1} \in \cdot|Y_m,m < 2n-1) = H_{D_n \cup F}(Y_{2n-1},\cdot)$ and $P(Y_{2n} \in \cdot|Y_m, m < 2n) = H_{D_n}(Y_{2n-1},\cdot)$. Note $(Y_{2n},n \geq 0)$ is just the chain $(Z_n,n \geq 0)$ defined above; hence Y_{2n} converges a.s. and Y_{2n} has distribution $H_{D_n}(x,\cdot)$. We prove in the next lemma that Y_m converges a.s. Assuming that lemma the convergence (2.6) is proved as follows. Let $\Lambda_1 = \{Y_m$ converges to a point in $(F-\Delta)^o\}$ and $\Lambda_2 = \{Y_m$ converges to a point in $(F-\Delta)^c\}$ (A^c denotes the complement of A). Then since $H_{D_n}(x,\cdot)$ converges to $H_D(x,\cdot)$ weakly and $H_D(x,\partial(F-\Delta)) = 0$, $P(\Lambda_1 \cup \Lambda_2) = 1$. By Lemma 2.7(ii), $H_D(Y_{2n},F-\Delta) \to 1$ a.s. on Λ_1; therefore

$$(2.8) \quad \lim_n \int H_D(Y_{2n},dz)H_F(z,F-\Delta) = 1 = \lim_n H_F(Y_{2n},F-\Delta) \text{ a.s. on } \Lambda_1.$$

The fact that Y_m converges a.s. implies $\overline{\lim_k} H_{D_k \cup F}(Y_{2n},F-\Delta) \to 0$ in probability on Λ_2. Using H4) and the fact $H_{D \cup F}(Y_{2n},\partial(F-\Delta)) = 0$ a.s. which holds because $P[H_D(Y_{2n},\partial(F-\Delta))] = H_D(x,\partial(F-\Delta)) = 0$ and so $H_D(Y_{2n},\partial(F-\Delta)) = 0$ a.s., we have $H_{D \cup F}(Y_{2n},F-\Delta) \to 0$ in probability on Λ_2. Now if $H_{D \cup F}(y,F-\Delta) < \varepsilon$ then the total variation of $H_{D \cup F}(y,dz) - H_D(y,dz)$ is less than 2ε (as always by H2)). Thus

$$\begin{aligned} &|H_F(Y_{2n},F-\Delta) - \int H_D(Y_{2n},dz)H_F(z,F-\Delta)| \\ &\quad = |\int H_{D \cup F}(Y_{2n},dz)H_F(z,F-\Delta) - \int H_D(Y_{2n},dz)H_F(z,F-\Delta)| \to 0 \end{aligned}$$

in probability on Λ_2. Combining this and (2.8) (and noting $P(Y_{2n} \in dy)$

$= H_{D_n}(x,dy))$ one has (2.6). It remains to establish

LEMMA 2.8[1]. Y_m *converges a.s.* dP.

PROOF. Assume $\Lambda = \{Y_m \text{ does not converge}\}$ has probability $a > 0$. Since Y_{2n} converges a.s. we have, a.s. on Λ, $Y_m \in F - \Delta$ i.o. and Y_{2n} converges to a point in $(F - \Delta)^c$. We may assume that for some $D_0 \in \mathcal{D}$ disjoint from $F - \Delta$, Y_{2n} converges to a point in D_0 a.s. on Λ. Denote by $[Y_m]^*$ the set of limit points of $\{Y_m\}$ and define

$$c(A) = P(\Lambda \cap \{[Y_m]^* \cap A \neq \emptyset\})$$

for open and compact $A \subset K$. The set function c is increasing, strongly subadditive and right continuous on the class of compact sets. Therefore by the Choquet capacity theorem c is extended from compact sets to a capacity c on all subsets A of K; c is continuous from below and for $B \in \mathcal{B}$

$$c(B) = \sup\{c(F_1): F_1 \subset B, F_1 \text{ is compact}\}.$$

Now for all n and $y \neq \Delta$ choose compact $V(n,y)$ (all from a countable collection of sets) such that each $V(n,y)$ is a neighborhood of y contained in $B(y,1/n)$ and $V(n,y)$ decreases in n, and $\int H_{D_0}(y,dz)H_{V(n,y)\cup\Delta}f(z)$ is in $b\mathcal{B}^n$ if f is. The last condition is possible because of H3), or rather its equivalent condition that $H_D f \in b\mathcal{B}^n$ for $f \in b\mathcal{B}^n$ (proved below in Theorem 2.11). Define

$$A_n = \{y \in F - \Delta: \int H_{D_0}(y,dz)H_{V(n,y)\cup\Delta}(z,V(n,y)) < 1 - 1/n\}.$$

Then by hypothesis H6) $A_n \uparrow F - \Delta$. Thus $c(A_n) \uparrow c(F - \Delta) = P(\Lambda) = a$.

---

[1] In the proof of this lemma we use H4); but it is valid assuming only H4A.1) in place of H4). See the arguments in the proof of Theorem 3.4 where the situation is similar; however the proof of Theorem 3.4 is quite a bit harder and will be given less than complete details.

Therefore $c(A_{n_0}) > a/2$ for some n_0. Since $A_{n_0} \in \mathcal{B}^n$, there exist B_1, B_2 in $\mathcal{B}$ with $B_1 \subset A_{n_0} \subset B_2$ such that $H_{C \cup \Delta}(x,C) = 0$ for all compact $C \subset B_2 - B_1$. Now there is a compact $F_0 \subset B_2 \cap (F - \Delta)$ with $c(F_0) = P(\Lambda \cap \{[Y_m]^* \cap F_0 \neq \emptyset\}) > a/2$. Let us consider the following refinement of $(Y_m, m \geq 0)$, still denoted $(Y_m, m \geq 0)$ (note (Y_m) is a refinement of (Z_n)):

$$P(Y_0 = x) = 1;\ P(Y_{3n-2} \in \cdot \mid Y_m, m < 3n-2) = H_{F \cup D_n}(Y_{3n-3}, \cdot);$$

$$P(Y_{3n-1} \in \cdot \mid Y_m, m < 3n-1) = H_{F_0 \cup D_n}(Y_{3n-2}, \cdot);$$

$$P(Y_{3n} \in \cdot \mid Y_m, m < 3n) = H_{D_n}(Y_{3n-1}, \cdot).$$

Note that the old (Y_m) is imbedded in the new (Y_m) as $(Y_0, Y_1, Y_3, Y_4, Y_6, \ldots)$. We claim that for this refinement (Y_m), $P(Y_m \in F_0 - B_1) = 0$ for all m. For if not then there exists compact $C \subset F_0 - B_1 \subset B_2 - B_1$ with $P(Y_m \in C) > 0$ for some m. But since $H_{C \cup \Delta}(x,C) = 0$, by repeatedly using (2.1) with $C \cup \Delta$ as the F and $F \cup D_n$, $F_0 \cup D_n$ and D_n as the D there one obtains $P(Y_m \in C) = 0$ for all m from the definition of (Y_m). Now if $Y_m = y \in F_0 \cap B_1 \subset A_{n_0}$, then the conditional probability that there exist $m_2 > m_1 > m$ with $Y_{m_1} \in D_0$ and $Y_{m_2} \in F_0 \cap V(n_0,y)$ is less than $1/n_0$; this again follows from repeated applications of (2.1) (with $(F_0 \cap V(n_0,y)) \cup \Delta$ as the F there) and the definition of A_{n_0}. From the compactness of F_0 it follows that $Y_m \in K - F_0$ for all sufficiently large m, a.s. on Λ (the same Λ defined from the old (Y_m), now as $(Y_0, Y_1, Y_3, \ldots)$). By shifting the initial position $Y_0 = x$ to a suitable point Y_{m_0} we may assume $P(Y_m \in F_0$ for some $m)$ $< a/4$. Now it is an easy argument, using the chain (Y_m), H2) and H4) to define a sequence of integer $n_k \uparrow \infty$, compact sets $C_k \downarrow F_0$ with $F_0 \subset C_k^{\circ}$ and a neighborhood U of F_0 with $\bar{U} \cap D_0 = \emptyset$ and $H_{D \cup F_0}(x, \partial U) = 0$ such that for all k

$$H_{C_k \cup D_{n_k}}(x,U) - H_{C_{k+1} \cup D_{n_k}}(x,U) > a/4.$$

This is impossible since both $H_{C_k \cup D_{n_k}}(x,\cdot)$ and $H_{C_{k+1} \cup D_{n_k}}(x,\cdot)$ converge weakly to $H_{D \cup F_0}(x,\cdot)$, so that the difference on the left side of the above converges to 0. This completes the proof of Lemma 2.8 and so that of Theorem 2.6. □

The proof of the following theorem also uses the above lemma.

THEOREM 2.9. *Under* H1), H2), H3) *and* H6), H4) *implies* H4A) *(and the existence of* ν_m *in the statement of* H4A.2)).

PROOF. Only H4A.2) needs proof. Let x be fixed, $D_n \downarrow D$, $F_k \uparrow K-D$. Let $F = F_{k \cup \Delta}$ for a fixed k. Define (Y_m) as in Lemma 2.8 from D_n and F. Then by Lemma 2.7 Y_m converges a.s. Since $F-\Delta$ and D_n are disjoint if n is sufficiently large, $Y_m \notin F-\Delta$ for all sufficiently large m. Applying Lemma 2.7(ii) with D_n and x there replaced by $D_n \cup F$ and points Y_{2n_0} for n_0 sufficiently large we have the following: for $\delta > 0$, $H_{D \cup F}(Y_{2n}, B(Y_{2n},\delta)) \to 1$ in probability. Since Y_{2n} has distribution $H_{D_n}(x,\cdot)$ and $H_{D_n}(x,\cdot)$ converges weakly to $H_D(x,\cdot)$, $H_{D_n}(x,dy)H_{D \cup F}(y,dz)$ converges weakly to $H_D(x,dy)\varepsilon_y(dz)$ which is then the measure $\nu_k(dy,dz)$ in H4A.2). Since ν_k is independent of k ($\nu_k\{(y,z): y = z\} = 1$ but this is not needed), H4A.2) follows. □

REMARK. We observe here that the measures ν_m in H4A.2) always exist under conditions of Theorem 2. As in the proof of Theorem 3.4 one can show that Lemma 2.8 is valid under conditions of Theorem 2 (see also the footnote about Lemma 2.8). Applying this fact to the same (Y_m) as in the above proof, and using the easy-to-establish fact that for $f \in b\mathcal{B}^*$, $H_{D \cup F}f(Z_n)$, $n \geq 0$, form a martingale, where (Z_n) is another supplementary chain and is as defined in Lemma 2.7 with D_n and x there replaced by $D_n \cup F$ and points Y_{2n_0}, n_0 sufficiently large, one can prove again $H_{D_n}(x,dy)H_{D \cup F}(y,dz)$ converges weakly, and our assertion follows.

Next we prove a fact that was mentioned in the discussion about H5) and also used in the proof of Theorem 2.1.

THEOREM 2.10. *Assume only* H1), H2), H3) *and* H6). *Fix* x *and an increasing sequence* D_n *and let* (W_n) *be as in* H5). *Then* (i) W_n *converges a.s.;* (ii) *for* $F \in \mathcal{D}$, $f \in b\mathcal{B}^*$, $H_F f(W_n)$ *converges a.s. on* $\{W_\infty \notin F\}$.

PROOF. The chain (W_n) is well-defined[1)] since by H2) its finite dimensional distributions are consistent. We prove (i) below. From this proof and the argument in Lemma 2.7 we will have $\overline{\lim_m} H_{D_n \cup F}(W_m, F) \to 0$ a.s. on $\{W_\infty \notin F\}$; (ii) follows with a martingale argument. To show (i) suppose $\Lambda = \{W_n$ does not converge$\}$ has positive measure. Note that for any m, $W_n \notin D_m$ for all sufficiently large n a.s. on Λ. Using the capacity argument in the proof of Lemma 2.8 we then have the following: there exist compact C_i, open U_i, $i = 1, 2$, and $\delta > 0$, $a > 0$ such that (1) $C_i \subset U_i$, $U_2 \subset K - U_1$; (2) for each $y \in C_1$ there is a compact neighborhood V(y) of y satisfying $\int H_{K-U_1}(y,dz)H_{V(y) \cup \Delta}(z,V(y)) < 1-\delta$; (3) the set $\{[W_n]^* \cap C_i \neq \emptyset$ for both $i = 1, 2\}$ has measure $a > 0$. It is possible to obtain compact F_i with $C_i \subset F_i \subset U_i$, $F_i - C_i \subset \bigcup_n D_n$ and $P(W_n \in F_i$ i.o. for both i$) > a$. Define another supplementary (but this time homogeneous) chain $(Y_m, m \geq 0)$ as follows:

[1)] The existence of (W_n) is justified as follows: (W_n) as a process in $(K,\mathcal{B})$ can be defined on the sample space $(K^\infty,\mathcal{B}^\infty) = (K\times K\times\cdots,\mathcal{B}\times\mathcal{B}\times\cdots)$. Now the completion of $\mathcal{B}^\infty$ contains $\mathcal{B}^*\times\mathcal{B}^*\times\cdots$ and (W_n) as a process in $(K,\mathcal{B}^*)$ is taken as defined on $(K^\infty,\mathcal{B}^*\times\mathcal{B}^*\times\cdots)$. Similar procedures justify the definition of Z in §3 and Z_∞ in §4 with appropriate product spaces as their sample spaces. Of course the above also applies to the existence of (Z_n) and (Y_m) in Lemmas 2.7 and 2.8; but their existence directly follows from I. Tulcea's theorem.

$$P(Y_0 \in \cdot) = H_{F_1 \cup F_2 \cup \Delta}(x,\cdot) \quad \text{and}$$

$$P(Y_{m+1} \in \cdot | Y_k, k \le m) = H_{F_2 \cup \Delta}(Y_m,\cdot) \quad \text{if} \quad Y_m \in F_1 \cup \Delta$$

$$= H_{F_1 \cup \Delta}(Y_m,\cdot) \quad \text{if} \quad Y_m \in F_2 \cup \Delta .$$

One can show that $P(Y_m \in C_i$ i.o. for both i) > a. But this contradicts the properties of C_i by arguments similar to those in the proof of Lemma 2.8. We omit the details. □

The results in the next theorem have been used in stating hypothesis H3) and proving Lemma 2.8 (and in a similar way Theorems 2.9 and 2.10).

THEOREM 2.11. *Assume only* H1) *and* H2). (i) *The family* $\mathcal{B}^n$ *of nearly Borel sets is a σ-algebra.* (ii) H3) *implies and so is equivalent to:* $H_D f \in b\mathcal{B}^n$ *for all* D *and* $f \in b\mathcal{B}^n$.

PROOF. For (i) we need only prove that if $B_n \in \mathcal{B}^n$, $n \ge 1$, then $B = \bigcup_n B_n \in \mathcal{B}^n$. Let μ be a given probability measure on $\mathcal{B}$; then there exist B_{ni}, $i = 1, 2$, in $\mathcal{B}$ with $B_{n1} \subset B_n \subset B_{n2}$ such that $H_{C_1 \cup \Delta}(\mu, C_1) \equiv \int \mu(dx) H_{C_1 \cup \Delta}(x, C_1) = 0$ for all compact $C_1 \subset B_{n2} - B_{n1}$. Let $B^i = \bigcup_n B_{ni}$; then $B^1 \subset B \subset B^2$ and we claim $H_{C \cup \Delta}(\mu, C) = 0$ for all compact $C \subset B^2 - B^1$. For otherwise there exists such C with $H_{C \cup \Delta}(\mu, C_1) > 0$ for some compact $C_1 \subset C \subset (B_{n2} - B_{n1})$ for some n. But by H2) and H1)

$$H_{C_1 \cup \Delta}(x, C_1) \ge \int_{C_1} H_{C \cup \Delta}(x, dy) H_{C_1 \cup \Delta}(y, C_1) = H_{C \cup \Delta}(x, C_1)$$

and we must have $H_{C_1 \cup \Delta}(\mu, C_1) > 0$. To prove (ii) we note first that the following criterion about $b\mathcal{B}^n$ is easy to establish: $f_0 \in b\mathcal{B}^n$ iff for any probability measure μ there exist f_1, f_2 in $b\mathcal{B}$ with $f_1 \le f_0 \le f_2$ such that $H_{C \cup \Delta}(\mu, C) = 0$ for all compact $C \subset \{f_2 > f_1\}$. Thus let μ be given; we must show there exist g_1, g_2 in $b\mathcal{B}$ with $g_1 \le H_D f \le g_2$ such that $H_{C \cup \Delta}(\mu, C) = 0$ for all compact $C \subset \{g_2 > g_1\}$. Let f_1, f_2 be in

$b\mathcal{B}$ with $f_1 \le f \le f_2$ such that $H_{V\cup\Delta}(\mu,V) = 0$ for all compact $V \subset B = \{f_2 > f_1\}$. Since by H3) $H_D f_i \in b\mathcal{B}^n$ there are g_i, g_i' in $b\mathcal{B}$ with $g_1 \le H_D f_1 \le g_1'$, $g_2' \le H_D f_2 \le g_2$ such that $H_{C\cup\Delta}(\mu,C) = 0$ for all compact $C \subset A_1 = \{g_1' > g_1\}$ or $C \subset A_2 = \{g_2 > g_2'\}$. Now $g_1 \le H_D f \le g_2$. To show $H_{C\cup\Delta}(\mu,C) = 0$ for all compact $C \subset \{g_2 > g_1\}$ we may assume $C \subset \{g_2' > g_1'\} \subset \{H_D f_2 > H_D f_1\} \equiv A$; this is because $H_{C\cup\Delta}(\mu,A_1 \cup A_2) = 0$ (see the argument in (i)). Of course $H_D(y,B) > 0$ for $y \in A$. Suppose $H_{C\cup\Delta}(\mu,C) > 0$ for some compact $C \subset A$; then $\int H_{C\cup\Delta}(\mu,dy)H_D(y,B) > 0$ and so $\int H_{C\cup\Delta}(\mu,dy)H_D(y,V) > 0$ for some compact $V \subset B \cap D$. Since $H_D(y,V) \le H_{V\cup\Delta}(y,V)$ we have $\int H_{C\cup\Delta}(\mu,dy)H_{V\cup\Delta}(y,V) > 0$. But $H_{V\cup\Delta}(y,V) = 0$ and consequently $H_{C\cup V\cup\Delta}(\mu,V) = 0$; it then follows from H2) that

$$H_{C\cup\Delta}(\mu,dy) = H_{C\cup V\cup\Delta}(\mu,dy).$$

We thus have the contradiction

$$\begin{aligned} H_{V\cup\Delta}(\mu,V) &= \int H_{C\cup V\cup\Delta}(\mu,dy)H_{V\cup\Delta}(y,V) \\ &= \int H_{C\cup\Delta}(\mu,dy)H_{V\cup\Delta}(y,V) > 0. \end{aligned}$$ □

3. Approximating Chains

In the next five sections we prove Theorem 2; thus we are given a family $\{H_D(x,\cdot): D \in \mathcal{D},\ x \in K\}$ of measures on K and a real-valued function e on K satisfying H1), H2), H3), H4A), H5) and H7). As in [4] we will define a sequence of (homogeneous) Markov processes on K whose time parameter ranges over ordinals less than a certain fixed ordinal. Such a process will be called a (Markov) chain here (it was called a generalized random walk in [4]). This sequence of chains admits a projective limit process to be defined in the next section, which contains all information about the trajectories of the sought-for right process. In this section we define a typical chain and establish its properties.

Consider a finite open covering $\mathfrak{U}_0$ of K closed under (finite) union and intersection. Let $\mathfrak{D}_0 = \{(K-U)\cup\Delta : U \in \mathfrak{U}_0\}$; then $\mathfrak{D}_0 \subset \mathfrak{D}$ and $\mathfrak{D}_0$ is closed under (finite) union and intersection. Let $U(x) = \bigcap\{U \in \mathfrak{U}_0 : x \in U\}$ and $D(x) = (K - U(x)) \cup \Delta$; so for $x \neq \Delta$, $D(x)$ is the largest $D \in \mathfrak{D}_0$ not containing x.

Let

$$Q(x,B) = H_{D(x)}(x,B).$$

$Q(x,B)$ is a Markov kernel on $(K,\mathcal{B}^*)$; note $Q(\Delta,\Delta) = 1$. Let π denote the countable ordinal ω^ω; here ω stands for the first infinite ordinal and $\omega^\omega = \lim_n \omega^n$. Let K_α be copies of K for $\alpha < \pi$ and $\mathfrak{W}$ be the product space $\prod_{\alpha<\pi} K_\alpha$; $Z_\alpha = Z(\alpha)$ denotes the α-th coordinate on $\mathfrak{W}$. With the Z_α regarded as measurable mappings into $(K,\mathcal{B}^*)$ let $\mathcal{H} = \sigma(Z_\alpha, \alpha < \pi)$, $\mathcal{H}_\alpha = \mathcal{H}(\alpha) = \sigma(Z_\beta, \beta \leq \alpha)$ and $\mathcal{H}_{\alpha-} = \mathcal{H}(\alpha-) = \sigma(Z_\beta, \beta < \alpha)$. If τ is a (Z_α)-stopping time (i.e. $(\mathcal{H}_\alpha)$-stopping time, of course τ assumes values $\alpha \leq \pi$), $\mathcal{H}(\tau)$ is defined in the usual way. The shift operators on $\mathfrak{W}$ are denoted by θ_α, which satisfy $Z_\beta \circ \theta_\alpha = Z_{\alpha+\beta}$ ($\alpha + \beta$ is the β-th ordinal after α). Now define a chain $Z = (Z_\alpha, \alpha < \pi;\ Q_x, x \in K)$ on $(K,\mathcal{B}^*)$ satisfying the following: for all x

$$Q_x(Z_0 = x) = 1,\quad Q_x(Z_{\alpha+1} \in B \mid \mathcal{H}_\alpha) = Q(Z_\alpha, B)$$

and for a limit ordinal (an ordinal $\neq 0$ and with no predecessor) α

$$Q_x(Z_\alpha \in \cdot \mid \mathcal{H}_{\alpha-}) = w\text{-}\lim_{\beta\uparrow\alpha} H_{D(\alpha)}(Z_\beta, \cdot) \equiv \nu_\alpha(\cdot)$$

where $D(\alpha) = \lim_{\beta\uparrow\alpha} \bigcap_{\beta<\gamma<\alpha} D(Z_\gamma)$ (note $D(\alpha) \in \mathfrak{D}_0$ and equals $\bigcap_{\beta<\gamma<\alpha} D(Z_\gamma)$ for all sufficiently large $\beta < \alpha$). The above weak limit will be shown to exist a.s. (i.e. a.s. Q_x for all x); let it stand for $\varepsilon_\Delta(\cdot)$ if it does not exist. In the sequel we denote by $\mathcal{L}$ the set of limit ordinals $\alpha < \pi$; also we will use the notation $D(\alpha + 1) = D(Z_\alpha)$ for all $\alpha < \pi$.

The following lemmas are proved together by induction on $\alpha < \pi$. Define, for $A \in \mathcal{B}^*$, $\tau_A = \inf\{\beta < \pi : Z_\beta \in A\}$ with the convention $\inf \emptyset = \pi$ here; and $H_A^\alpha(x,B) = Q_x(Z(\tau_A) \in B, \tau_A < \alpha)$.

LEMMA 3.1. *For any* x, $D \in \mathcal{D}_0$ *and* $f \in b\mathcal{B}^*$, $H_D f(x) = Q_x(H_D f(Z_{\tau\wedge\alpha})]$ $= Q_x[f(Z_\tau);\tau \le \alpha] + Q_x[H_D f(Z_\alpha);\tau > \alpha]$ *where* $\tau = \tau_D$; *in particular* $H_D^\alpha(x,\cdot) \le H_D(x,\cdot)$, *for all* $\alpha < \pi$.

LEMMA 3.2. *For all* $\alpha \in \mathcal{L}$, (i) $\nu_\alpha(\cdot) \equiv w\text{-}\lim_{\beta\uparrow\alpha} H_{D(\alpha)}(Z_\beta,\cdot)$ *exists a.s.*; (ii) *for each* $f \in b\mathcal{B}^*$, $H_{D(\alpha)}f(Z_\beta) \to \int \nu_\alpha(dy)f(y)$ *a.s. and in the mean under any* Q_x *as* $\beta \uparrow \alpha$.

PROOF OF LEMMA 3.1. The one-step induction from α to $\alpha + 1$ follows from the fact that if $\tau > \alpha$ then $D \subset D(Z_\alpha) = D(\alpha + 1)$ so that $H_D f(Z_\alpha)$ $= H_{D(\alpha+1)}H_D f(Z_\alpha) = \int Q(Z_\alpha,dy)H_D f(y)$. To prove the equality at an $\alpha \in \mathcal{L}$ we apply Lemma 3.2 with the same α (whose proof relies on Lemma 3.1 at smaller ordinals). For $\beta < \alpha$ we have

$$H_D f(x) = Q_x[H_D f(Z_\tau);\tau \le \beta] + Q_x[H_D f(Z_\beta);\tau > \beta]. \tag{3.1}$$

If $\tau \ge \alpha$ then $D \subset D(\alpha)$ and so $H_D f(Z_\beta) = H_{D(\alpha)}H_D f(Z_\beta)$ for $\beta < \alpha$; therefore by Lemma 3.2(ii) with f replaced by $H_D f$ the second term on the right of (3.1) converges to

$$Q_x[\int \nu_\alpha(dy)H_D f(y);\tau \ge \alpha] = Q_x[H_D f(Z_\alpha);\tau \ge \alpha]$$

as $\beta \uparrow \alpha$. The desired equality for this α follows. □

PROOF OF LEMMA 3.2. For $D \in \mathcal{D}_0$, $f \in b\mathcal{B}^*$ and with $\tau = \tau_D$, $\{H_D f(Z_{\tau\wedge\beta}), \mathcal{H}(\tau \wedge \beta), \beta < \alpha\}$ is a martingale under any Q_x; this follows from Lemma 3.1, assumed to hold for all ordinals smaller than α and the Markov property of $(Z_\beta, \beta < \alpha; Q_y, y \in K)$. Thus $H_D f(Z_{\tau\wedge\beta})$ converges a.s. Q_x. Let f run through a countable dense subset of $\mathbf{C}(K)$ and D through all sets in $\mathfrak{D}_0$ and applying the Riesz representation theorem we have $\nu_{\alpha,D}(\cdot) \equiv w\text{-}\lim_{\beta\uparrow\alpha} H_D(Z_{\tau_D \wedge \beta},\cdot)$ exists for all $D \in \mathfrak{D}_0$ a.s. Hence a.s. on $\{\tau_{D(\alpha)} = \alpha\}$, $\nu_\alpha(\cdot) = \nu_{\alpha,D(\alpha)}(\cdot) = w\text{-}\lim_{\beta\uparrow\alpha} H_{D(\alpha)}(Z_\beta,\cdot)$ exists. Now as $\gamma \uparrow \alpha$, $Q_x[\tau_{D(\alpha-\gamma)} \circ \theta_\gamma = \alpha - \gamma] \to 1$ ($\alpha - \gamma$ is the ordinal α' satisfying $\gamma + \alpha' = \alpha$); thus (i) follows from the above and the Markov property of $(Z_\beta, \beta < \alpha)$. To prove (ii) note that for $f \in b\mathcal{B}^*$ and $D \in \mathfrak{D}_0$, with the following $\mathcal{H}(\tau) \wedge \mathcal{H}(\alpha-)$-measurable function (again $\tau = \tau_D$)

$$H_D f(Z_\tau)\cdot 1_{[\tau<\alpha]} + 1_{[\tau \geq \alpha]}\cdot \int \nu_{\alpha,D}(dy)f(y) \tag{3.2}$$

attached at the end of the martingale $\{H_D f(Z_{\tau\wedge\beta}), \mathcal{H}(\tau \wedge \beta), \beta < \alpha\}$ under any Q_x, we have an expanded martingale. This is true for $f \in \mathcal{C}(K)$ by the definition of $\nu_{\alpha,D}$, and therefore true for all $f \in b\mathcal{B}^*$. Again by the martingale convergence theorem $H_D f(Z_{\tau\wedge\beta})$ converges a.s. and in the mean to (3.2) under any Q_x. Thus $H_D f(Z_\beta) \to \int \nu_{\alpha,D}(dy)f(y)$ a.s. and in the mean on $\{\tau_D \geq \alpha\}$ under Q_x as $\beta \uparrow \alpha$. This being true for all $D \in \mathfrak{D}_0$, $H_{D(\alpha)} f(Z_\beta) \to \int \nu_\alpha(dy)f(y)$ a.s. and in the mean on $\{\tau_{D(\alpha)} = \alpha\}$ under any Q_x as $\beta \uparrow \alpha$. As in (i), (ii) follows from this and the Markov property.□

The existence of $Z = (Z_\alpha, \alpha < \pi;\ Q_x, x \in K)$ on the sample space $(\mathcal{W},\mathcal{H})$ is justified by induction and the Kolmogorov extension theorem, with the trivial completion procedure mentioned in the footnote in the proof of Theorem 2.10.

THEOREM 3.3. *Z satisfies the following:* (i) *A.s.* $Z_\alpha = \Delta$ *for all sufficiently large* $\alpha < \pi$; *indeed, if* $m \geq |\mathfrak{D}_0|$ *(the cardinality of* $\mathfrak{D}_0$*) then a.s.* $Z_\alpha = \Delta$ *for all* $\alpha \geq \omega^m$. (ii) *For all* x *and* $D \in \mathfrak{D}_0$, $Q_x[Z(\tau_D) \in \cdot\,] = H_D(x,\cdot)$.

PROOF. For $\alpha \in \mathcal{L}$, $Z(\alpha) \in D(\alpha) = \lim_{\beta\uparrow\alpha} \bigcap_{\beta<\gamma<\alpha} D(Z_\gamma)$ a.s. So a.s. $Z(\omega)$ is in the intersection of at least two distinct $D \in \mathfrak{D}_0$ if $Z(\alpha) = \Delta$ for no $\alpha < \omega$, and so are $Z(n\omega)$ for all n. It follows that a.s.

$Z(\omega^2)$ is in the intersection of at least three distinct $D \in \mathcal{D}_0$ if $Z(\alpha) = \Delta$ for no $\alpha < \omega^2$, and so are $Z(n\omega^2)$ for all n. In general a.s. $Z(n\omega^k)$ is in the intersection of $k+1$ distinct $D \in \mathcal{D}_0$ except if $Z(\alpha) = \Delta$ for some $\alpha < n\omega^k$ (in which case $Z(\gamma) = \Delta$ for all $\gamma > \alpha$ a.s.). Thus (i) follows. (ii) now follows from Lemma 3.1 by taking an $\alpha > \omega^m$ where $m = |\mathcal{D}_0|$, since $\tau_D < \alpha$ a.s. □

THEOREM 3.4. *A.s. the path $\alpha \to Z_\alpha$ has left limits (at all $\alpha \in \mathcal{L}$).*

PROOF. Suppose for some x and $\alpha \in \mathcal{L}$, Z_β does not converge as $\beta \uparrow \alpha$ on a set of positive Q_x-measure. Then by the argument used in the proof of Lemma 2.8 one obtains the following: there exist $a > 0$, $\delta > 0$, compact C_i and F_i, $i = 1,2$ with $C_i \subset F_i^\circ$, $F_1 \cap F_2 = \emptyset$, $C_1 \subset \{e_{F_2 \cup \Delta} > \delta\}$, $C_2 \subset \{e_{F_1 \cup \Delta} > \delta\}$ and

$$Q_x([Z_\beta]^* \cap C_i \neq \emptyset \quad \text{for both } i = 1,2) > a \tag{3.3}$$

where $[Z_\beta]^*$ denotes the set of limit points of Z_β as $\beta \uparrow \alpha$, i.e. the set of $z = \lim_k Z(\beta_k)$ for some sequence $\beta_k \uparrow \alpha$. Here we have used condition H7.1A) (see the discussion about H7.1) in §1) — in contrast H6) was used in Lemma 2.8 to get a contradiction. We now define a refinement $Z^1 = (Z^1_\gamma, \gamma < \pi; Q_x, x \in K)$ of Z (Z^1 is a chain that has Z imbedded in it, and the imbedding justifies the same Q_x notation) as follows: Let $\mathcal{D}_1$ be the minimal family of sets containing $\mathcal{D}_0$ and $C_i \cup \Delta$, $F_i \cup \Delta$, $i = 1,2$, and closed under union and intersection. Then let Z^1 be the chain defined from $\mathcal{D}_1$ (i.e. from the measures $H_D(x,\cdot)$, $x \in K$, $D \in \mathcal{D}_1$) in the same manner as Z from $\mathcal{D}_0$. Since $\mathcal{D}_0 \subset \mathcal{D}_1$, so that the sets $D(x)$ in the definition of Z are in $\mathcal{D}_1$, from Theorem 3.3(ii) applied to Z^1 we see that Z is clearly imbedded in Z^1 (there exist (Z^1_α)-stopping times $0 = \sigma_0 \leq \sigma_1 \leq \cdots \leq \sigma_\alpha \leq \cdots$, $\alpha < \pi$ such that $(Z^1(\sigma_\alpha), \alpha < \pi)$ is equivalent to, and identified as, $(Z_\alpha, \alpha < \pi)$). (See the beginning of §4 for a more systematic discussion.) Let $\tau_1, \tau_2, \ldots$ be successive times of alternate visits to $C_1 \cup \Delta$ and $F_2 \cup \Delta$ by (Z^1_γ), i.e. with $\tau_0 = 0$

$$\tau_{2n-1} = \inf\{\gamma > \tau_{2n-2} : Z^1_\gamma \in C_1 \cup \Delta\}, \ \tau_{2n} = \inf\{\gamma > \tau_{2n-1} : Z^1_\gamma \in F_2 \cup \Delta\}.$$

We show $Z^1(\tau_n) = \Delta$ for sufficiently large n a.s. For by Theorem 3.3 (ii) applied to Z^1 and alternately to $D = C_1 \cup \Delta$ and $D = F_2 \cup \Delta$

$$\begin{aligned} e(x) &\geq H_{C_1\cup\Delta} e(x) = Q_x[e(Z^1_{\tau_1})] = Q_x[e_{F_2\cup\Delta}(Z^1_{\tau_1}) + H_{F_1\cup\Delta} e(Z^1_{\tau_1})] \\ &\geq \delta\, Q_x[Z^1_{\tau_1} \neq \Delta] + Q_x[e(Z^1_{\tau_2})] \\ &= \text{1st term} + Q_x[e_{C_1\cup\Delta}(Z^1_{\tau_2}) + H_{C_1\cup\Delta} e(Z^1_{\tau_2})] \\ &\geq \text{1st term} + Q_x[e(Z^1_{\tau_3})] \geq \cdots \\ &\geq \text{1st term} + \delta\, Q_x[Z^1_{\tau_3} \neq \Delta] + Q_x[e(Z^1_{\tau_5})] \geq \cdots \\ &\geq n\,\delta\, Q_x[Z^1_{\tau_{2n-1}} \neq \Delta] \end{aligned}$$

for all n; thus $Q_x[Z^1(\tau_n) \neq \Delta] \to 0$. This fact of course also holds with C_1, F_2 replaced by C_2, F_1 respectively. Now by shifting x to a suitable point $Z^1_{\gamma_0}$ we may assume $Q_x(\Gamma) < a/2$ where $\Gamma = \{Z^1_\gamma \in C_1 \cup C_2$ for some $\gamma\}$, without affecting the validity of (3.3) (of course Z being considered as imbedded in Z^1). Denote the set on the left-hand side of (3.3) by Λ. Then $Q_x(\Lambda - \Gamma) > a/2$. We will define a sequence of compact sets D_n decreasing to $C_1 \cup C_2 \cup \Delta$ such that $H_{D_n}(x,\cdot)$ does not converge weakly, thus obtaining a contradiction to condition H4A.1) of H4A; specifically for a neighborhood U of C_1 whose closure is disjoint from $C_2 \cup \Delta$ (note C_1 and $C_2 \cup \Delta$ are disjoint) we will have $\underline{\lim}_n |H_{D_n}(x,U) - H_{D_{n+1}}(x,U)| > a/4$. To do this we claim that there exist disjoint compact sets C_m, $m \geq 3$, satisfying

(1) $C_{2k+i} \subset F^\circ_i \cap B(C_i, 1/k) \cap C^c_i$, $i = 1,2$ (recall $B(A,\delta) = \{y : d(y,A) < \delta\}$),

(2) $Q_x[\Lambda - \Gamma, \bigcap_{m=3}^{\infty} \{Z_\beta \in C_m$ for some $\beta\}] > a/4$,

(3) $\inf\{e_{F_2 \cup \Delta}(y) : y \in C_{2k+1}\} > 0$ and $\inf\{e_{F_1 \cup \Delta}(y) : y \in C_{2k+2}\} > 0$,

(4) $\sup\{H_{C_3 \cup C_5 \cup \cdots \cup C_{2k-1} \cup \Delta}(y; C_3 \cup C_5 \cup \cdots \cup C_{2k-1}) : y \in C_{2k}\} < 1/k$,

$\sup\{H_{C_4 \cup C_6 \cup \cdots \cup C_{2k} \cup \Delta}(y, C_4 \cup C_6 \cup \cdots \cup C_{2k}) : y \in C_{2k+1}\} < 1/k$.

The C_m are defined successively. It is easy to see that (1), (2), (3) can be achieved. To achieve (4) one considers refinements Z^m, $m \geq 2$, of Z^1 where Z^m is defined from the following family $\mathfrak{D}_m$ in the same way as Z^1 from $\mathfrak{D}_1$: $\mathfrak{D}_m$ is the minimal family of sets containing $\mathfrak{D}_1$ and $C_j \cup \Delta$, $3 \leq j \leq m+1$, and closed under union and intersection. That C_m, $m \geq 5$, can be obtained satisfying (4) follows from considering (3) and the chain Z^{m-2}. Now for $1 \leq j_1 < j_2 < j_3$ with j_1, j_3 odd and j_2 even, let $C^i = \bigcup_{k > j_i} C_{2k+i}$ and $D^i = C^i \cup C_1 \cup C_2 \cup \Delta$, $i = 1,2,3$. Let $\hat{Z}$ be the refinement of Z^1 with $\mathfrak{D}_1$ replaced by the minimal family of sets containing $\mathfrak{D}_1$ and D^i, $i = 1,2,3$, and closed under union and intersection. We claim that for fixed j_1, if j_2 is sufficiently larger than j_1 and j_3 sufficiently larger than j_2,

$$H_{D^1 \cup D^2}(x, D^1) - H_{D^2 \cup D^3}(x, D^3) > a/4.$$

This follows from an easy argument by considering $\hat{Z}$ and using (2) and (4) above. Now we can define a sequence $D_n \downarrow C_1 \cup C_2 \cup \Delta$ such that each D_{2n-1} is a set like $D^1 \cup D^2$ and each D_{2n} like $D^2 \cup D^3$ above, and for all n

$$|H_{D_n}(x, F_1^\circ) - H_{D_{n+1}}(x, F_1^\circ)| > a/4.$$

Since $U = F_1^\circ$ is a neighborhood of C_1 whose closure is disjoint from $C_2 \cup \Delta$ we have a contradiction to H4A.1). □

In the rest of the section we define in Z a family of positive functions indexed by α and increasing with α that will be used to approximate a family of times canonical in the sought-for right process.

DEFINITION. Define $e(0) = 0$, $e(\alpha+1) = e_{D(\alpha+1)}(Z_\alpha)$ (recall $D(\alpha+1) = D(Z_\alpha)$), and for $\alpha \in \mathfrak{L}$ or $\alpha = \pi$, $e(\alpha) = \lim_{\beta\uparrow\alpha} e_{D(\alpha)}(Z_\beta)$ (note $e(\alpha) = 0$ for all sufficiently large $\alpha \le \pi$); then define

$$R(\alpha) = \sum_{\beta\le\alpha} e(\beta), \qquad \alpha \le \pi.$$

LEMMA 3.5. *For* $\alpha \in \mathfrak{L}$, $e_{D(\alpha)}(Z_\beta)$ *converges a.s. and in the mean to* $e(\alpha)$ *under any* Q_x *as* $\beta \uparrow \alpha$.

PROOF. By the arguments in the proof of Lemma 3.2, it suffices to show that for any $D \in \mathcal{D}_0$, $\{e_D(Z_{\tau\wedge\beta}), \mathcal{H}(\tau\wedge\beta), \beta < \alpha\}$, where $\tau = \tau_D$, is a uniformly integrable supermartingale under Q_x. The supermartingale defining inequality at two successive times β and $\beta+1$ is immediate from the inequality $e_D(y) \ge H_{D(y)}e_D(y)$ if $y \notin D$, i.e. $D \subset D(y)$ (compare the proof of Lemma 3.1). For the inequality at $\beta' < \beta \in \mathfrak{L}$ we claim first that

$$\begin{aligned}\varliminf_{\gamma\uparrow\beta} e_D(Z_\gamma) &= \varliminf_{\gamma\uparrow\beta} (e_{D(\beta)}(Z_\gamma) + H_{D(\beta)}e_D(Z_\gamma)) \\ &\ge \varliminf_{\gamma\uparrow\beta} H_{D(\beta)}e_D(Z_\gamma) \ge \int \nu_\beta(dy)e_D(y)\end{aligned}$$

a.s. Q_x on $\{\tau \ge \beta\}$; this follows from applying Lemma 3.2(ii) to $f = e_D \wedge n$ and then letting $n \to \infty$. Using Fatou's lemma we then have $Q_x[e_D(Z_{\tau\wedge\beta})|\mathcal{H}(\tau\wedge\beta')] \le e_D(Z_{\tau\wedge\beta'})$. To prove the desired uniform inequality it suffices to show $\{e(Z_\beta), \beta < \pi\}$ is uniformly integrable under Q_x, since $e_D \le e$ and $e_D(Z_\tau) = 0$. Suppose this uniform integrability fails to hold; then for some $\varepsilon > 0$ and all $a > 0$ there exists $F \in \mathcal{D}$ with $F - \Delta \subset \{e \ge a\}$ such that $Q_x[e(Z_\alpha);\ Z_\alpha \in F \cup \Delta] > \varepsilon$. Consider the refinement $Z' = (Z'_\gamma, Q_x)$ of Z constructed with $\mathcal{D}_0$ replaced by the minimal family $\mathcal{D}'$ containing $\mathcal{D}_0$ and F and closed under union and intersection. Let $\sigma = \inf\{\gamma : Z'_\gamma \in F\}$ and σ' be the time in Z' corresponding to the time α in Z (now imbedded in Z'); these are Z'-stopping times. Since $\{e(Z'_\gamma), \gamma < \pi\}$ is a supermartingale by the above established fact (noting $e = e_{\{\Delta\}}$ and $e(\Delta) = 0$), $Q_x[e(Z'_\sigma)] \geq Q_x[Z'_{\sigma\vee\sigma'})]$. Clearly the last expression is at least $Q_x[e(Z_\alpha); Z_\alpha \in F - \Delta]$; on the other hand $Q_x[e(Z'_\alpha)] = H_F e(x)$ by Theorem 3.3(ii) applied to Z'. So $H_F e(x) \geq \varepsilon$. Since a is arbitrary, we have a contradiction to condition H7.4). □

LEMMA 3.6. *For* $\alpha \in \mathfrak{L}$, $H_{D(\alpha)}e(Z_\beta) \to \int \nu_\alpha(dy)e(y)$ *a.s. and in the mean under any* Q_x *as* $\beta \uparrow \alpha$.

PROOF. Let $f_n = e \cdot 1_{[e \geq n]}$ and $g_n = e - f_n$. Then by Lemma 3.2(ii) $H_{D(\alpha)}f_n(Z_\beta)$ converges a.s. and in the mean to $\int \nu_\alpha(dy)f_n(y)$ under Q_x as $\beta \uparrow \alpha$. To show the desired mean convergence it suffices to show $\sup_{\beta<\alpha} Q_x[H_{D(\alpha)}g_n(Z_\beta)] \to 0$ as $n \to \infty$. But $Q_x[H_D g_n(Z_\beta)] = Q_x[g_n(Z(\tau_D)\circ\theta_\beta)]$, which by the argument in the proof of Lemma 3.5 proving the uniform integrability is small uniformly in β and $D \in \mathfrak{D}_0$ if n is large; therefore the above supremum indeed converges to 0. The desired a.s. convergence will be seen in the next proof (of course not needed there). □

THEOREM 3.7. *For all* x, $e(x) = Q_x[R(\pi)]$ *and* $e_D(x) = Q_x[R(\tau_D)]$ *for* $D \in \mathfrak{D}_0$.

PROOF. Although the second assertion can be proved directly in the same manner as the first, we show it is an immediate consequence of the latter and Theorem 3.3(ii). For with $\tau = \tau_D$

$$e(x) = Q_x[R(\pi)] = Q_x[R(\tau) + R(\pi)\circ\theta_\tau]$$
$$= Q_x[R(\tau)] + Q_x[Q_{Z(\tau)}(R(\pi))] = Q_x[R(\pi)] + H_D e(x)$$

so that $Q_x[R(\pi)] = e(x) - H_D e(x) = e_D(x)$. To prove the first assertion, it suffices to show $\{R(\alpha) + e(Z_\alpha), \mathcal{H}(\alpha), \alpha < \pi\}$ is a martingale under Q_x; for if $\alpha \geq \omega^m$ where $m = |\mathfrak{D}_0|$ then $R(\alpha) + e(Z_\alpha) = R(\pi)$. The martingale equality at two successive times α and $\alpha+1$ follows from $e(y) = e_{D(y)} + H_{D(y)}e(y)$. To prove the equality at $\alpha' < \alpha \in \mathcal{L}$, we have as $\beta \uparrow \alpha$

$$e(Z_\beta) = e_{D(\alpha)}(Z_\beta) + H_{D(\alpha)}e(Z_\beta) \to e(\alpha) + \int \nu_\alpha(dy)e(y)$$

in the mean under Q_x by Lemmas 3.5 and 3.6. It follows that

$$Q_x[R(\alpha) + e(Z_\alpha)|\mathcal{H}(\alpha')] = Q_x[R(\alpha) + \int \nu_\alpha(dy)e(y)|\mathcal{H}(\alpha')] = R(\alpha') + e(Z_{\alpha'}),$$

completing the proof. Let us now observe that the a.s. convergence in Lemma 3.6 follows immediately from the a.s. convergence of $e(Z_\beta)$ (as the difference of the positive martingale $\{R(\beta) + e(Z_\beta)\}$ and the increasing $R(\beta)$) and that of $e_{D(\alpha)}(Z_\beta)$ (by Lemma 3.5). □

4. The projective limit process Z_∞

We now define a sequence of chains Z_n on the state space $(K,\mathcal{B}^*)$ similar to the chain Z defined in §3, such that for each n, Z_n has a natural imbedding in Z_{n+1}, and consequently the sequence admits a projective limit process Z_∞. Consider a sequence of finite open coverings $\mathfrak{U}_n$, $n \geq 1$, of K satisfying:

(i) $\mathfrak{U}_n$ is closed under (finite) union and intersection,

(ii) $\mathfrak{U}_n \subset \mathfrak{U}_{n+1}$,

(iii) for all x the set $U_{nx} = U(n,x) \equiv \bigcap\{U \in \mathfrak{U}_n : x \in U\}$ has diameter $< 1/n$.

Let $\mathcal{D}_n = \{(K-U)\cup\Delta : U \in \mathfrak{U}_n\}$, $\mathcal{D}_\infty = \bigcup_n \mathcal{D}_n$, and $D(n,x) = D_{nx} = (K-U_{nx}) \cup \Delta$. Then $\mathcal{D}_n$ is closed under (finite) union and intersection; $\mathcal{D}_n \uparrow \mathcal{D}_\infty \subset \mathcal{D}$; for $x \neq \Delta$, $D(n,x)$ is the largest set in $\mathcal{D}_n$ not containing x, and if $y \notin D(n,x)$ then $d(x,y) < 1/n$ and $D(n,x) \subset D(k,y)$ for $k > n$.

Recall $(\mathfrak{W},\mathcal{H})$ is the product space of $(K_\alpha,\mathcal{B}^*_\alpha) \equiv (K,\mathcal{B}^*)$, $\alpha < \pi$; Z_α is the αth coordinate on $\mathfrak{W}$. Define for each n a chain $Z_n = (Z_{n\alpha}, \alpha < \pi;\ Q_x^{(n)}, x \in K)$ on the sample space $(\mathfrak{W},\mathcal{H})$ in the same way as Z in §3 with the family $\mathcal{D}_n$ playing the role of $\mathcal{D}_0$ (there is a temporary ambiguity about the notation Z_n as a chain and as a random variable Z_α when $\alpha = n$; but this will soon be cleared away below); i.e. Z_n has one-step transition probability $Q_n(y,B) = H_{D(n,y)}(y,B)$ and for $\alpha \in \mathcal{L}$

$$Q_x^{(n)}(Z_\alpha \in \cdot \,|\, \mathcal{H}(\alpha-)) = w - \lim_{\beta\uparrow\alpha} H_{D(n,\alpha)}(Z_\beta,\cdot)$$

where $D(n,\alpha) = \lim_{\beta\uparrow\alpha} \bigcap_{\beta<\gamma<\alpha} D(n,Z_\gamma)$.

The property of such a sequence of chains Z_n that makes possible the embedding of Z_n in Z_k, $k > n$, is the following:

$$(4.1) \qquad Q_x^{(k)}[Z(\tau_D) \in \cdot\,] = H_D(x,\cdot),\ x \in K,\ D \in \mathcal{D}_n,\ k \ge n.$$

This of course follows from Theorem 3.3(ii) applied to Z_k. We now proceed to define the projective limit process Z_∞ of the Z_n.

DEFINITION. For each n let $\sigma_{n\alpha}$, $\alpha < \pi$, be the (Z_α)-stopping times defined as follows: first set $\sigma_n = \tau_{D(n,y)}$ if $Z_0 = y$; then define $\sigma_{n0} = 0$, $\sigma_{n,\alpha+1} = \sigma_{n\alpha} + \sigma_n \circ \theta(\sigma_{n\alpha})$ (recall $\theta(\beta) = \theta_\beta$ are the shift operators) with the understanding $\sigma_{n,\alpha+1} = \pi$ if $\sigma_{n\alpha} = \pi$, and for $\alpha \in \mathcal{L}$

$$\sigma_{n\alpha} = \lim_{\beta\uparrow\alpha}[\alpha_{n\beta} + \tau_{D(n,\alpha)} \circ \theta(\sigma_{n\beta})]$$

(note this is not necessarily $\sup_{\beta<\alpha} \sigma_{n\beta} = \lim_{\beta\uparrow\alpha} \sigma_{n\beta}$). These stopping times are only considered in Z_k for $k \ge n$.

DEFINITION. Let $\mathfrak{W}_n = \{w \in \mathfrak{W}\colon \sigma_{n\alpha}(w) = \alpha \text{ for all } \alpha < \pi\}$, $\Omega^\circ = \prod_n \mathfrak{W}_n$, π_n be the projection of Ω° to $\mathfrak{W}_n$, and $Z_{n\alpha} = Z(n,\alpha) = Z_\alpha \circ \pi_n$. For a (Z_α)-stopping time τ, let $\tau^{(n)} = \tau \circ \pi_n$; $\tau^{(n)}$ is often simply denoted τ when misunderstanding is unlikely, and in particular $Z(n,\tau)$ means

$Z(n,\tau^{(n)})$.

DEFINITION. Let

$$\Omega = \{\omega \in \Omega^\circ : Z(n+1,\sigma_{n\alpha})(\omega) = Z(n,\alpha)(\omega) \text{ for all } n \geq 1,\ \alpha < \pi\}$$

and define on Ω the following σ-algebras:

$$\mathcal{H}_{n\alpha} = \mathcal{H}(n,\alpha) = \sigma(Z_{n\beta},\ \beta \leq \alpha),\quad \mathcal{H}_n = \mathcal{H}(n) = \sigma(Z_{n\alpha},\ \alpha < \pi),$$

$$\mathcal{G} = \bigvee_n \mathcal{H}_n = \sigma(Z_{n\alpha},\ n \geq 1,\ \alpha < \pi),$$

of course with $Z_{n\alpha}$ considered as random variables in $(K,\mathcal{B}^*)$.

Now $Q_x^{(n)}(\mathbb{W}_n) = 1$ for all x; $\pi_n : \Omega \to \mathbb{W}_n$ is easily seen to be onto. Thus for each n, $(Z_{n\alpha}; Q_x^{(n)}\pi_n)$ is equivalent to $(Z_\alpha; Q_x)$. Obviously $\mathcal{H}_n \subset \mathcal{H}_{n+1}$.

THEOREM 4.1. *For all* x *and* n, $Q_x^{(n)} \circ \pi_n = Q_x^{(n+1)} \circ \pi_{n+1}$ *on* $\mathcal{H}_n$.

PROOF. We prove that the two measures agree on $\mathcal{H}(n,\alpha)$ by induction on α. The one-step induction follows from $Q_y^{(n+1)}[Z(\sigma_n) \in \cdot] = H_{D(n,y)}(y,\cdot) = Q_y^{(n)}[Z(1) \in \cdot]$, a consequence of (4.1). The induction at an $\alpha \in \mathcal{L}$ follows from the definition of the distributions of Z_n and Z_{n+1} at limiting ordinals and the fact $D(n,\alpha)\circ\pi_n = D(n+1,\sigma_{n\alpha})\circ\pi_{n+1}$ which implies

$$w\text{-}\lim_{\beta \uparrow \alpha} H_{D(n,\alpha)\circ\pi_n}(Z(n,\beta),\cdot) = w\text{-}\lim_{\beta\uparrow\alpha} H_{D(n+1,\sigma_{n\alpha})\circ\pi_{n+1}}(Z(n+1,\sigma_{n\beta}),\cdot). \quad \square$$

Now for each x there exists a finitely additive P^x on $\bigcup_n \mathcal{H}_n$ that extends $Q_x^{(n)}\circ\pi_n$ for all n. Since Ω is a Borel measurable set of the product $\mathbb{W}\times\mathbb{W}\times \cdots$ (or of $\prod_n \mathbb{W}_n$) of compact spaces, using the Kolmogorov extension theorem and then the completion procedure in the footnote in Theorem 2.10 we extend each P^x as a measure on $\mathcal{G} = \sigma(Z_{n\alpha}, n \geq 1,\ \alpha < \pi)$. The process

$$Z_\infty = (Z_{n\alpha},\ n \geq 1,\ \alpha < \pi; P^x,\ x \in K)$$

on the state space $(K,\mathcal{B}^*)$ is the projective limit of the sequence of chains Z_n. For a fixed n, $(Z_{n\alpha}, \alpha < \pi; P^x, x \in K)$ is of course equivalent to Z_n; from now on the notation Z_n is used to denote this (new) chain. The Markov property of the Z_n gives rise to a Markov property of Z_∞; to state it we need the following definition.

DEFINITION. For a $(Z_n, \alpha < \pi)$-stopping time σ, $\mathfrak{H}(n,\sigma)$ is of course defined in the usual way. For $k \geq n$, let $\sigma(k,n,\alpha)$ denote the $(Z_{k\gamma}, \gamma < \pi)$ stopping time $\sigma^{(k)}_{n\alpha}$. We define the σ-algebras

$$\mathfrak{G}_{n\alpha} = \mathfrak{G}(n,\alpha) = \bigvee_{k \geq n} \mathfrak{H}(k,\sigma(k,n,\alpha)).$$

If τ is a stopping time relative to $(\mathfrak{G}_{n\alpha}, \alpha < \pi)$, $\mathfrak{G}_{n\tau} = \mathfrak{G}(n,\tau)$ is again defined in the usual way. The shift operators $\theta_{n\alpha} = \theta(n,\alpha)$ on Ω are defined by requiring $Z(k,\gamma) \circ \theta_{n\alpha} = Z(k,\sigma(k,n,\alpha) + \gamma)$ for all $k \geq n$, $\gamma < \pi$.

The Markov property of Z_∞ can now be stated as follows: for any x, stopping time τ relative to $(\mathfrak{G}_{n\alpha}, \alpha < \pi)$, and $\eta \in b\mathfrak{G}$

$$(4.2) \qquad P^x(\eta \circ \theta_{n\tau} | \mathfrak{G}(n,\tau)) = P^{Z(n,\tau)}(\eta) \quad \text{on } \{\tau < \pi\}.$$

This follows easily from the (strong) Markov property of the chains Z_n.

Let us restate (4.1) for future reference:

$$(4.3) \qquad P^x(Z(k,\tau_D) \in \cdot) = H_D(x,\cdot), \; x \in K, \; D \in \mathfrak{D}_n, \; k \geq n,$$

here τ_D of course stands for $\tau^{(k)}_D = \inf\{\alpha : Z(k,\alpha) \in D\}$; note incidentally that although the value of τ_D varies with k the position $Z(k,\tau_D)$ does not — so long as $k \geq n$).

The following definition emphasizes the (random) ordering of the "time" set of Z_∞.

DEFINITION. Denote $\mathfrak{T} = \{(n,\alpha) : n \geq 1, \alpha < \pi\}$; we introduce the following ordering on $\mathfrak{T}$ dependent on $\omega \in \Omega$ (elements of $\mathfrak{T}$ are considered as functions of Ω for this ordering): (m,β) is less than (resp. equal to) (n,α), where $m \le n$, if $\sigma(n,m,\beta)$ is less than (resp. equal to) α.

The following trivial fact is recorded for future reference.

THEOREM 4.2. *Almost surely the "path"* $(n,\alpha) \to Z(n,\alpha)$ *on* $\mathfrak{T}$ *has no oscillation, and is right continuous (with the obvious understanding that* $(n_j,\alpha_j) \downarrow (n_0,\alpha_0)$ *means for any* $(n,\alpha) > (n_0,\alpha_0)$, $(n_0,\alpha_0) \le (n_j,\alpha_j) < (n,\alpha)$ *for all sufficiently large* j*).*

PROOF. The first assertion is immediate from Theorem 3.4; the second from the fact that if $(n,\gamma) \le (n_j,\alpha_j) < (n,\gamma+1)$ then $Z(n_j,\alpha_j) \in U(n,Z_{n\gamma})$ and so its distance from $Z(n,\gamma)$ is less than $1/n$.□

Next we examine the approximating times associated with the Z_n.

DEFINITION. Let $D(n,\alpha+1) = D(n,Z_{n\alpha})$ (recall $D(n,y)$ is the largest $D \in \mathfrak{D}_n$ not containing y if $y \ne \Delta$), $D(n,\alpha) = \lim_{\beta\uparrow\alpha} \bigcap_{\beta<\gamma<\alpha} D(n,Z_\gamma)$ (this notation has been used in defining the original Z_n) for $\alpha \in \mathfrak{L}$ or $\alpha = \pi$; $e_{n,\alpha+1} = e(n,\alpha+1) = e_{D(n,\alpha+1)}(Z_{n\alpha})$, $e_{n\alpha} = e(n,\alpha) = \lim_{\beta\uparrow\alpha} e_{D(n,\alpha)}(Z_\beta)$ for $\alpha \in \mathfrak{L}$ or $\alpha = \pi$. Then define

$$R_{n\alpha} = R(n,\alpha) = \sum_{\beta\le\alpha} e(n,\beta), \quad \alpha \le \pi \ ; \quad \text{and}$$

$$R(n,m,\beta) = R(n,\sigma(n,m,\beta)) \quad \text{for } n \ge m,\ \beta < \pi.$$

From Theorem 3.6 we have for all x and $D \in \mathfrak{D}_n$

$$(4.4) \qquad P^x(R(n,\pi)) = P^x(R(n,\tau_\Delta)) = e(x);\ P^x(R(n,\tau_D)) = e_D(x).$$

(Again τ_D in the above means $\tau_D^{(n)}$.) In §5 we study the convergence of $R(n,m,\beta)$ for fixed m,β; for the moment we prove an important preliminary fact.

THEOREM 4.3. $\{R(n,\pi), n \ge 1\}$ *is uniformly integrable under any* P^x.

PROOF. Suppose the contrary; then there exist x and $\varepsilon > 0$ such that for all $b > 0$ one can find n with $P^x[R(n,\pi); R(n,\pi) > b] > \varepsilon$. Fixing b and such n define $\tau = \tau_n = \inf\{\alpha < \pi: R(n,\alpha) \geq b/3$ or $e(Z_{n\alpha}) \geq b/3\}$ (again $\inf \emptyset = \pi$ here). From the definition of $e(\alpha)$ and $R(n,\alpha)$ it is clear that $\{R(n,\pi) > b\} \subset \{\tau < \pi\}$ and $R(n,\tau) < 2b/3$ on $\{\tau < \pi\}$. Now using (4.4) and Markov property

$$(4.5)\quad P^x(e(Z_{n\tau});\tau<\pi) = P^x(R(n,\pi) - R(n,\tau);\tau < \pi)$$

$$\geq P^x(R(n,\pi) - R(n,\tau);R(n,\pi)> b) \geq P^x(R(n,\pi)/3;R(n,\pi)> b) > \varepsilon/3.$$

Also

$$e(x) \geq P^x(R(n,\pi);\tau < \pi)$$

$$\geq P^x(R(n,\pi);\tau<\pi,R(n,\tau)\geq b/3) + P^x(R(n,\pi) - R(n,\tau);\tau<\pi,R(n,\tau) < b/3)$$

$$\geq \frac{b}{3} P^x(\tau<\pi,R(n,\tau)\geq b/3) + P^x(e(Z_{n\tau});\tau < \pi,R(n,\tau) < b/3)$$

$$\geq \frac{b}{3} P^x(\tau < \pi),$$

so that $P^x(\tau < \pi) \leq 3e(x)/b$. For a given $a > 0$ choose b such that $P^x(\tau<\pi) < \varepsilon/6a$; then by (4.5) $P^x(e(Z_{n\tau});\tau<\pi,e(Z_{n\tau}) \geq a) > \varepsilon/3 - \varepsilon/6 = \varepsilon/6$. Using an argument similar to the one in the proof of Lemma 3.5 one can then find $D \in \mathcal{D}$ with $D - \Delta \subset \{e > a\}$ such that $H_D e(x) > \varepsilon/3$. Since a is arbitrary this contradicts H7.4). □

In a number of proofs in the sequel we need to consider refinements of Z_∞ defined as follows.

DEFINITION. Let $F_1,\ldots,F_k$ be sets in $\mathcal{D}$. The $\{F_1,\ldots,F_k\}$-refinement $\tilde{Z}_\infty$ of Z_∞ is the projective limit process constructed in the same way as Z_∞ but with the collection $\{\mathcal{D}_n, n \geq 1\}$ replaced by $\{\tilde{\mathcal{D}}_n, n \geq 1\}$ where $\tilde{\mathcal{D}}_n$ is the minimal family of sets containing $\mathcal{D}_n$ and $\{F_1,\ldots,F_k\}$ and closed under union and intersection. This $\tilde{Z}_\infty$ will be denoted $(\tilde{Z}_{n\alpha}, n \geq 1, \alpha < \pi; P^x, x \in K)$.

As observed in §3, each $Z_n = (\tilde{Z}_{n\alpha},\alpha < \pi;P^x,x \in K)$ is a refinement of Z_n; therefore $\tilde{Z}_\infty$ is a refinement of Z_∞ (Z_∞ is imbedded in $\tilde{Z}_\infty$), and this justifies the use of the same notation P^x for the measures.

5. Convergence of Approximating Times

In this section we define random times $S_{m\beta} = S(m,\beta)$ in Z_∞ which together with the trajectories $(m,\beta) \to Z(m,\beta)$ determine the sought-for process (X_t). If (X_t) has no holding points, these are the stopping times in (X_t) corresponding to the "times" $\sigma_{m\beta}$ in Z_∞. To this end we study the convergence of $R(n,m,\beta)$ as $n \to \infty$. Under condition (NH) below the detailed proofs will be given, while under condition (H) the proofs will be somewhat sketchy. Before stating these conditions recall that in §1 we have defined $x \neq \Delta$ to be a holding point if $h(x) \equiv \inf\{e_D(x): x \notin D\} > 0$. Let H denote the set of holding points.

Condition (NH): No holding points are present, i.e. $H = \emptyset$.

Condition (H): Holding points are present, i.e. $H \neq \emptyset$.

Under condition (NH), for any x the function $R(n,m,\beta)$ converges in P^x-measure as $n \to \infty$, and the convergence is uniform (in P^x-measure) in β; a sequence $\{n_k\}$ is then defined, whose values depend on the initial position $Z(m,0)$, such that $R(n_k,m,\beta)$ converges a.s. (a.s. P^x for all x) for all pairs (m,β). The limits are then the functions $S(m,\beta)$. Under condition (H), we must modify $R(n,m,\beta)$ before establishing the convergence in measure. As mentioned in §1, we claimed in [4] that the functions $R(n,m,\beta)$ (denoted $\xi(m,\beta,n)$ in §6 of [4]), $n \geq m$, form a martingale; but it was false, caused by an oversight. (In [4] the corresponding projective limit process Z_∞ was only used as a tool in the construction of (X_t), which was defined from its transition semigroup.) Here the mistake is corrected.

DEFINITION. For $m \leq n$ let $e(n,m,0) = 0$ and for $\beta > 0$

$$e(n,m,\beta) = R(n,m,\beta) - R(n,m,\beta-)$$

with the understanding that $\beta-$ stands for $\beta - 1$ if the predecessor

$\beta - 1$ of β exists and $R(n,m,\beta-) = \sup_{\beta_1<\beta} R(n,m,\beta_1)$ if $\beta \in \mathfrak{L}$. Note we can write

$$e(n,m,\beta) = \sum\{e(n,\alpha): (m,\beta - 1) < (n,\alpha) \leq (m,\beta)\}$$

if $\beta - 1$ exists (recall the definition of the ordering of (n,α) in §4), and if $\beta \in \mathfrak{L}$

$$e(n,m,\beta) = \sum\{e(n,\alpha): (m,\beta-) \leq (n,\alpha) \leq (m,\beta)\}$$

where $(m,\beta-) \leq (n,\alpha)$ means $\alpha \geq \sup_{\gamma<\beta} \sigma(n,m,\gamma)$, i.e. $(n,\alpha) > (m,\gamma)$ for all $\gamma < \beta$.

LEMMA 5.1. *Assume condition* (NH). *For all* x *and* $\varepsilon > 0$

$$\lim_m \sup_{n\geq m} P^x(\sup_\beta e(n,m,\beta) > \varepsilon) = 0.$$

The lemma will be proved after the following theorem.

THEOREM 5.2. *Assume condition* (NH). *For any* x *and* m, $R(n,m,\beta)$ *converges as* $n \to \infty$ *uniformly in* P^x*-measure for* $\beta < \pi$.

PROOF. For notational simplicity we prove $R(n,\pi)$ converges in P^x-measure; it will be clear from the proof that the assertion of the theorem is valid. Thus we show that for $c > 0$

$$\sup_{n\geq m} P^x(|\sum_\beta (e(n,m,\beta) - e(m,\beta))| > c) \to 0 \tag{5.1}$$

as $m \to \infty$, since $R(m,\pi) = \sum_\beta e(m,\beta)$ and $R(n,\pi) = \sum_\beta e(n,m,\beta)$. Fix $\varepsilon > 0$ and $m < n$ (m,β are now free) define functions

$$V_\beta = e(n,m,\beta), \quad V'_\beta = V_\beta \cdot 1_{[V_\beta \leq \varepsilon]}, \quad V''_\beta = V_\beta - V'_\beta;$$

$$W_\beta = P^x(V_\beta | \mathfrak{G}(m,\beta-)) = e(m,\beta),$$

$$W'_\beta = P^x(V'_\beta | \mathfrak{G}(m,\beta-)), \quad W''_\beta = W_\beta - W'_\beta$$

where $\mathfrak{G}(m,\beta-) = \{\emptyset,\Omega\}$ if $\beta = 0$; $= \mathfrak{G}(m,\beta - 1)$ if $\beta - 1$ exists; $= \bigvee_{\gamma<\beta} \mathfrak{G}(m,\gamma)$ if $\beta \in \mathfrak{L}$ (see the definition of $\mathfrak{G}(m,\gamma)$ preceding the

Markov property (4.2)). That $P^x(V_\beta | \mathcal{G}(m,\beta-)) = e(m,\beta)$ follows from (4.4) and (4.2) and, in the case $\beta \in \mathcal{L}$, an easy argument based on Lemma 3.5. By Lemma 5.1 $\sup_{n\geq m} P^x(V''_\beta \neq 0 \text{ for some } \beta) \to 0$ as $m \to \infty$. Thus by the uniform integrability of $R(n,\pi)$ (Theorem 4.3) $\sup_{n\geq m} P^x(\sum_\beta V''_\beta) \to 0$, and so of course $\sup_{n\geq m} P^x(\sum_\beta W''_\beta) \to 0$. Hence to prove (5.1) it suffices to show

$$\sup_{n\geq m} P^x(|\sum_\beta (V'_\beta - W'_\beta)| > c) \to 0$$

as $m \to \infty$. By Chebyshev's inequality and the fact V'_β, W'_β are $\mathcal{G}(m,\beta_1-)$-measurable if $\beta < \beta_1$, we have

$$P^x(|\sum_\beta (V'_\beta - W'_\beta)| > c) \leq c^{-2} \sum_\beta P^x[(V'_\beta - W'_\beta)^2]$$

$$+ 2c^{-2} \sum_\beta \sum_{\beta_1>\beta} P^x[(V'_\beta - W'_\beta)P^x(V'_{\beta_1} - W'_{\beta_1} | \mathcal{G}(m,\beta_1-))]$$

provided the first sum is finite. The second sum vanishes since the conditional expectations are equal to 0. The first sum is

$$\leq c^{-2} \sum_\beta 4\varepsilon P^x(V'_\beta) \leq 4\varepsilon c^{-2} P^x(R(n,\pi)) = 4\varepsilon c^{-2}\, e(x),$$

the last equality by (4.4). Since ε is arbitrary, the proof is complete. □

PROOF OF LEMMA 5.1. Suppose the contrary; then for some x and $\varepsilon > 0$ there exist positive integers $m_1 < n_1 < \cdots < m_j < n_j < \cdots$ such that

$$(5.2) \qquad P^x(\sup_\beta e(n_j, m_j, \beta) > \varepsilon) > \varepsilon \quad \text{for all } j.$$

We show that (5.2) leads to a contradiction. Define for $m \leq n$, $\alpha < \pi$

$$(5.3) \qquad h(n,\alpha,m) = e_{D(m,\beta)}(Z_{n\alpha})$$

where β is such that $(m,\beta-) \leq (n,\alpha) < (m,\beta)$ (again $\beta-$ stands for $\beta - 1$ if it exists). We first prove two sublemmas.

SUBLEMMA 5.1.1. *Under* (5.2) *there exists* $c > 0$ *such that for all* j

$$P^x(\sup_\alpha h(n_j,\alpha,m_j) > c) > c.$$

PROOF. Fix $m = m_j$, $n = n_j$. Let $0 < c < \varepsilon/3a$ where $a > 1$ is to be determined. Define

$$\tau = \inf\{\alpha : R(n,\alpha) - R(n,m,\beta-) \geq \varepsilon/3 \text{ where } (m,\beta-) \leq (n,\alpha) < (m,\beta)\}$$

($\tau = \pi$ if no such α exists). It is clear that if $\tau < \pi$ and $h(n,\tau,m) \leq c$ then $R(n,m,\beta) - R(n,m,\beta-) > \varepsilon$ (where β satisfies $(m,\beta-) \leq (n,\alpha) < (m,\beta)$) with conditional probability $< 3c/\varepsilon < 1/a$. By considering (finitely many) successive iterates of τ and using the integrability of $R(n,\pi)$ we see that if a is sufficiently large then $P^x(\sup_\alpha h(n,\alpha,m) > c) > c$. By the uniform integrability of $R(n,\pi)$ under P^x (Theorem 4.3), $c > 0$ can be obtained so that the inequality holds for all $m = m_j$, $n = n_j$. □

SUBLEMMA 5.1.2. *Under* (5.2) *there exist* $c > 0$, *positive integers* $m_1 < n_1 < \cdots < m_j < n_j < \cdots$ *(where* $\{m_j\}$ *(resp.* $\{n_j\}$*) is a subsequence of the* m_j *(resp.* n_j*) in* (5.2)), *and for each* j *a* Z_{n_j}*-stopping time* τ_j *such that* $(n_j,\tau_j) \leq (n_{j+1},\tau_{j+1})$ *(using convention* $(n,\alpha) \leq (n',\pi)$ *for all* $\alpha \leq \pi$ *if* $n \leq n'$*) and* $(n_{j+1},\tau_{j+1}) < (n_j,\tau_j + 1)$ *if* $\tau_{j+1} < \pi$, *and*

$$P^x(\tau_j < \pi \text{ and } h(n_j,\tau_j,m_j) > c \text{ for all } j) > c.$$

PROOF. By the preceding sublemma $P^x(\sup_\alpha h(n_j,\alpha,m_j) > c$ for infinitely many $j) > c$ (the same c as in there; m_j, n_j are as in (5.2)). Since $h(n_j,\alpha,m_j) \leq h(n_j,\alpha,m_i)$ for $i < j$ by the definition of $h(n,\alpha,m)$, one can choose successively $m_1 < n_1 < \cdots < m_j < n_j < \cdots$ from the original m_j, n_j respectively such that

$$P^x(\sup_\alpha h(n_j,\alpha,m_j) > c \text{ for all } j) > c.$$

Define $\tau_j = \inf\{\alpha: h(n_j,\alpha,m_j) > c\}$. Then $P^x(\tau_j < \pi$ for all $j) > c$. Using (4.4) and the Markov property (4.2) one can easily show that a.s. P^x on $\{\tau_j < \pi$ for all $j\}$ the ordering property of (n_j,τ_j) asserted holds for sufficiently large j. By shifting the sequence m_j, n_j (and redefining τ_j if necessary) we then have the conclusion of the sublemma.□

To continue the proof of Lemma 5.1, let $Y = \lim_j Z(n_j,\tau_j)$ on $\Lambda = \{\tau_j < \pi$ and $h(n_j,\tau_j,m_j) > c$ for all $j\}$ where c, m_j, n_j, τ_j are as in Sublemma 5.1.2. Y exists a.s. by Theorem 4.2. For each y and $k \geq 1$ choose $C(k,y) \in \mathcal{D}$ (from a countable subfamily of $\mathcal{D}$) such that $e_{C(k,y)}(y)$ is $\mathcal{B}^*$-measurable and for $y \neq \Delta$

$$B(y,2^{-k-1}) - \Delta \subset K - C(k,y) \subset B(y,2^{-k}).$$

Since $e_{C(k,y)}(y) \downarrow h(y) \equiv \inf\{e_F(y): y \notin F\}$ for $y \neq \Delta$ and $h(y) \equiv 0$ by condition (NH), we have $A_k = \{y: e_{C(k,y)}(y) < c/2\} \uparrow K$ as $k \to \infty$. Fix k with $P^x(\Lambda, Y \in A_k) > c$ (possible since $P^x(\Lambda) > c$); then choose $D \in \mathcal{D}$ with $D \subset A_k$ (note $\Delta \in A_k$) and $P^x(\Lambda_1) > c$ where $\Lambda_1 = \{\Lambda, Y \in D\}$. Consider the $\{D\}$-refinement $\tilde{Z}_\infty$ of Z_∞ defined at the end of §4. The rest of the proof will be a bit sketchy but it is easy to fill in the details. The "time" in $\tilde{Z}_\infty$ corresponding to (n,α) in Z_∞ is denoted $(n,\tilde{\alpha})$ (a better notation may be $(n,\tilde{\alpha}(n))$) so that $\tilde{Z}(n,\tilde{\alpha}) = Z(n,\alpha)$; thus the Z_{n_j}-stopping time τ_j has value $\tilde{\tau}_{n_j}$ in $\tilde{Z}_{n_j}$ and Λ can be written as $\{\tilde{\tau}_j < \pi$ and $h(n_j,\tilde{\tau}_j,m_j) > c$ for all $j\}$ (note here h is not $\tilde{h}$). We claim that, a.s. P^x on Λ_1, there are no visits to D by $\tilde{Z}_\infty = (\tilde{Z}_{n\gamma})$ between successive times $(n_j,\tilde{\tau}_j)$ and $(n_{j+1},\tilde{\tau}_{j+1})$ for all sufficiently large j, i.e. for some j_0, $\tilde{Z}_{n\gamma} \notin D$ for $(n_{j_0},\tilde{\tau}_{j_0}) \leq (n,\gamma) \leq (n_j,\tilde{\tau}_j)$ for all $j > j_0$. This follows from applying (4.3), (4.4) and the Markov property to $\tilde{Z}_\infty$ and using the following fact: for $y \in D$ and $D_1 \in \mathcal{D}$, since $e_{C(k,y)}(y) < \frac{c}{2}$, $H_{C(k,y)\cup D_1}(y, A - C(k,y)) < 1/2$ where $A = \{z: e_{C(k,y)}(z) > c\}$; for from these we obtain that, if $n_j > 2^{k+1}$ and $j' > j$

$$P^x(\Lambda_1, \tilde{Z}_{n\gamma} \in D \text{ for some } (n,\gamma) \text{ with } (n_j,\tilde{\tau}_j) \le (n,\gamma) \le (n_{j'},\gamma_{j'})) < 1/2.$$

Consequently, by shifting x to some point $Z(n_{j_0},\tau_{j_0}) = \tilde{Z}(n_{j_0},\tilde{\tau}_{j_0})$ we may assume $\tilde{Z}_{n\gamma} \notin D$ for all $(n,\gamma) \le (n_j,\tau_j)$ for all j, a.s. P^x on Λ_1 (still with $P^x(\Lambda_1) > c$) excluding a subset of P^x-measure $< c/2$. Also, if $e_{D_1}(z) > c$ and $\operatorname{diam}(K - D_1) < 2^{-k-1}$ so that for $y \in K - D_1$ we have $C(k,y) \subset D_1$, then $e_{D_1 \cup D}(z) > c/2$; from this it follows that a.s. P^x on Λ_1, $\tilde{h}(n_j,\tilde{\tau}_j,m_j) > c/2$ for all j with $n_j > 2^{k+1}$. Now define D_n to be the smallest set in $\mathcal{D}_n$ containing D in its interior and $F_m = K - \bigcup\{U(m,y)\colon y \in D\}$ (see the beginning of §4 for the definition of $U(m,y)$). Then $D_n \downarrow D$ and $F_m \uparrow D^c$. It is not difficult to prove that for each m if n is sufficiently large

$$P^x(e_{F_m \cup D}(\tilde{Z}(n,\tau_{D_n})) \ge (c/2)^2 = c^2/4$$

where $\tau_{D_n} = \inf\{\gamma\colon \tilde{Z}_{n\gamma} \in D_n\}$. Since the left-hand side of the above is $\int H_{D_n}(x,dy)e_{F_m \cup D}(y)$ by (4.3) and (4.4), we have a contradiction to condition H7.3) of hypothesis H7). Lemma 5.1 is finally proved. □

DEFINITION. Assume condition (NH). For each x define $n_k = n_k(x)$ with $1 = n_0 < n_1 < \cdots < n_k < \cdots$ inductively by

$$n_k = \inf\{n > n_{k-1}\colon \sup_{n'>n} \sup_{m\le k} \sup_{\beta} P^x(|R(n,m,\beta) - R(n',m,\beta)| \ge 2^{-k}) \ge 2^{-k}\}.$$

(The existence of the n_k follows from Theorem 5.2.) Now for each pair (m,β) and $k \ge m$ define

$$R_k(m,\beta) = R(n_k,m,\beta)$$

where $n_k = n_k(Z(m,0))$ (i.e. $n_k = n_k(x)$ if the path starts at x); then define

$$S_{m\beta} = S(m,\beta) = \varliminf_{k} R_k(m,\beta).$$

THEOREM 5.3. *Assume condition* (NH). *The functions* $S(m,\beta)$, $m \geq 1$, $\beta < \pi$, *defined above satisfy the following:* (i) $S(m,\beta)$ *is* $\mathfrak{G}((m,\beta)-)$[1]-*measurable;* $S(m,0) = 0$; $S(m,\beta) = S(m',\beta')$ *if* $(m,\beta) = (m',\beta')$ *so that* $(m,\beta) \to S(m,\beta)$ *is well-defined as a function on* $\mathfrak{T}$ *and is increasing.* (ii) $R_k(m,\beta) \to S(m,\beta)$ *a.s.* (iii) *For all* x, $P^x(S(m,\beta)) = P^x(R(m,\beta))$, *in particular* $P^x(S(m,1)) = e_{D(m,x)}(x)$; *also, for* $D \in \mathfrak{D}_m$, $P^x(S(m,\tau_D)) = e_D(x)$. (iv) *(Additivity). A.s. for all* β, β', $S(m,\beta+\beta') = S(m,\beta) + S(m,\beta')\circ\theta_{m\beta}$. (v) *Define* $\mathfrak{S}(\omega) = \{S_{m\beta}(\omega)\colon m \geq 1,\ \beta < \pi\}$ *and* $S_\Delta(\omega) = \sup \mathfrak{S}(\omega)$; *then* $S_\Delta(\omega) < \infty$ *and* $\mathfrak{S}(\omega)$ *is dense in* $[0,S_\Delta(\omega)]$ *a.s.*

PROOF. (i) and (ii) need no proof. (iii) follows from the uniform integrability of $R(n,\pi)$ under P^x, the fact $P^x(R(n,m,\beta)) = P^x(R(m,\beta))$ for $n \geq m$, and (4.3). To prove (iv) we observe that for each y if an arbitrary sequence N_k of positive integers increases fast enough then $R(N_k,m,\beta')$ converges a.s. P^y and the limit is $S(m,\beta')$ a.s. P^y; thus for any probability measure $\nu(dy)$, $R(N_k,m,\beta') \to S(m,\beta')$ a.s. $P^\nu \equiv \int \nu(dy)P^y$ if $N_k \to \infty$ fast enough. Choose a subsequence N_k of $n_k(x)$ increasing fast enough; applying the above to $\nu(dy) = P^x(Z(m,\beta) \in dy)$ we have $R(N_k,m,\beta')\circ\theta_{m\beta} \to S(m,\beta')\circ\theta_{m\beta}$ a.s. P^x. (iv) then follows from $R(n,m,\beta+\beta') = R(n,m,\beta) + R(n,m,\beta')\circ\theta_{m\beta}$. For (v), $S_\Delta < \infty$ a.s. follows from $P^x(S_\Delta) = e(x)$ and the denseness property is an easy consequence of Lemma 5.1. □

Next we define functions $S(m,\beta)$ under condition (H). In this case the functions $R(n,m,\beta)$ have to be modified. The proofs will only be sketched. First we need

DEFINITION. (Assume condition (H).) Denoting $\mathfrak{T}(m,\beta) = \{(n,\alpha)\colon n \geq m,\ (m,\beta-) \leq (n,\alpha) < (m,\beta)\}$, we define for $\varepsilon > 0$

---

[1] $\mathfrak{G}((m,\beta)-)$ is the minimal σ-algebra containing $\mathfrak{G}(n,\alpha) \cap \{(n,\alpha) < (m,\beta)\}$ for all (n,α); recall the σ-algebras $\mathfrak{G}(n,\alpha)$ were defined preceding the Markov property (4.2).

$$\Lambda(m,\beta,\varepsilon) = \{\text{for all } m' \geq m \text{ there is } n > m' \text{ such that } h(n,\alpha,m') > \varepsilon \text{ for some } \alpha \text{ with } (n,\alpha) \in \mathfrak{T}(m,\beta)\},$$

$$\Lambda(m,\beta) = \bigcup_{\varepsilon > 0} \Lambda(m,\beta,\varepsilon);$$

$$\Gamma(n,m,\beta,\varepsilon) = \{h(n,\alpha,m) > \varepsilon \text{ for some } \alpha \text{ with } (n,\alpha) \in \mathfrak{T}(m,\beta)\},$$

$$\Gamma(m,\beta,\varepsilon) = \bigcup_{n \geq m} \Gamma(n,m,\beta,\varepsilon).$$

Note $\Lambda(m,\beta,\varepsilon) \uparrow \Lambda(m,\beta)$ as $\varepsilon \downarrow 0$ and $\Gamma(n,m,\beta,\varepsilon) \uparrow \Gamma(m,\beta,\varepsilon)$ as $n \to \infty$.

LEMMA 5.4 *(Assume condition* (H).*) For any* x, $c > 0$, $\delta > 0$ *there exists* $\varepsilon > 0$ *such that for all* $m < n$

$$P^x(\bigcup_\beta [\{e(n,m,\beta) > c\} - \Gamma(n,m,\beta,\varepsilon)]) < \delta.$$

PROOF. Use an argument similar to the one in the proof of Sublemma 5.1.2. □

DEFINITION. (Assume condition (H).) For a fixed x define integers $m_j = m_j(x)$, $n_j = n_j(x)$ with $1 = n_0 < m_1 < n_1 < \cdots\ m_j < n_j < \cdots$ by

$$m_j = \inf\{m > n_{j-1}:\ P^x(\bigcup_\beta [\Lambda(m,\beta,2^{-j})\Delta\Gamma(m,\beta,2^{-j})]) < 2^{-j}\},$$

$$n_j = \inf\{n > m_j:\ P^x(\bigcup_\beta [\Lambda(m_j,\beta,2^{-j})\Delta\Gamma(n,m_j,\beta,2^{-j})]) < 2^{-j}\}$$

(here Δ stands for symmetric difference).

It is easy to see the existence of m_j and n_j.

DEFINITION. (Assume condition (H).) Fix (m,β). With $m_j = m_j(Z(m,0))$, $n_j = n_j(Z(m,0))$, define a function $\bar{R}(n_j,m,\beta)$ where $m_j \geq m$ as follows. First, denoting for a fixed β_1

$$\gamma = \inf\{\alpha:\ (m_j,\beta_1-) \leq (n,\alpha) < (m_j,\beta_1) \text{ and } h(n,\alpha,m_j) > 2^{-j}, \text{ or } (n,\alpha) = (m_j,\beta_1)\},$$

we define

$$\bar{e}(n_j,m_j,\beta_1) = \sum\{e(n_j,\alpha): (m_j,\beta_1-) < (n_j,\alpha) \le (n_j,\gamma)\} + h(n_j,\gamma,m_j)\}$$

$$\text{if } (n_j,\gamma) < (m_j,\beta_1)$$

$$= e(n_j,m_j,\beta_1) \text{ otherwise.}$$

Then define

$$\bar{R}(n_j,m,\beta) = \sum\{\bar{e}(n_j,m_j,\beta_1): (m_j,\beta_1) \le (m,\beta)\}.$$

THEOREM 5.5. *(Assume condition* (H).*) For fixed* x *and* m, $\bar{R}(n_j,m,\beta)$ *converges in* P^x*-measure as* $j \to \infty$, *and the convergence in* P^x*-measure is uniform in* β.

PROOF (Sketch). Let $m_j = m_j(x)$, $n_j = n_j(x)$. Suppose τ_j is a Z_{n_j}-stopping time for each j such that (n_j,τ_j) is increasing. Denote by β_j the (random) ordinal satisfying $(n_j,\tau_j) \in \mathfrak{T}(m_j,\beta_j)$ if $\tau_j < \pi$. Then for $c > 0$ we have $(n_{j+1},\tau_{j+1}) \in \mathfrak{T}(m_j,\beta_j)$ for all sufficiently large j a.s. P^x on $\Lambda = \{\tau_j < \pi$ for all j, $\underline{\lim}_j\, h(n_j,\tau_j,m_j) > c\}$, and by a (super-)martingale argument one can show $h(n_j,\tau_j,m_j)$ converges a.s. P^x on Λ. The proof of the theorem uses this fact, Lemma 5.4, and an argument similar to the proof of Theorem 5.2. □

DEFINITION. (Assume condition (H).) For a fixed x let $n_{j(k)} = n_{j(k)}(x)$ be a subsequence of $n_j = n_j(x)$ such that $n_{j(1)} = n_1$ and for $k > 1$

$$n_{j(k)} = \inf\{n_j > n_{j(k-1)}: \sup_{m\le k}\sup_{\beta}\sup_{i>j} P^x(|\bar{R}(n_j,m,\beta) - \bar{R}(n_i,m,\beta)| > 2^{-k}) < 2^k\}.$$

(Such a sequence exists by Theorem 5.5.) Then define

$$R_k(m,\beta) = \bar{R}(n_{j(k)},m,\beta) \quad \text{where} \quad n_{j(k)} = n_{j(k)}(Z(m,0));$$

$$S(m,\beta) = S_{m\beta} = \underline{\lim}_k\, R_k(m,\beta).$$

THEOREM 5.6. *Assume condition* (H). *Then statements* (i), (ii), (iii) *and* (iv) *of Theorem* 5.3 *hold for the functions* $S(m,\beta)$ *defined above.*

PROOF. Only (iv) needs proof; but it is similar to the corresponding argument in Theorem 5.3.

Statement (v) of Theorem 5.3 definitely fails to hold — typically $\mathfrak{S}(\omega)$ (defined from the above $S(m,\beta)$) is not dense in $[0, S_\Delta(\omega)]$, although we still have $P^x(S_\Delta) = e(x)$. □

6. Proof of $S(m,1) > 0$

In this section we prove a crucial property of the times $S(m,\beta)$ under either condition (NH) or (H), that $S(m,1) > 0$ a.s. P^x for $x \neq \Delta$. From the additivity of the $S(m,\beta)$ (Theorem 5.3(iv) and Theorem 5.6(iv)), it then follows that a.s. $S(m,\beta)$ is strictly increasing in $(m,\beta) \in \mathfrak{T}$ until $Z(m,\beta) = \Delta$. Again, the proofs under condition (H) are more involved and will be somewhat sketchy.

THEOREM 6.1. *(Assume condition* (NH).*)* $P^x(S(m,1) > 0) = 1$ *for all* $x \neq \Delta$ *and* $m \geq 1$.

To prove the theorem, assume $m = 1$ for convenience. Fix $x \neq \Delta$ and let $D = D(1,x)$. We need

DEFINITION (under either condition (NH) or (H)). Let $u(y) = P^y(S(1,\tau_D) = 0)$ where $\tau_D = \tau_D^{(1)} = \inf\{\alpha\colon Z(1,\alpha) \in D\}$. (Note that $u = 1$ on D.) We must show $u(x) = 0$. Assume

$$u(x) = a > 0. \tag{6.1}$$

We will obtain (under condition (NH)) a contradiction to condition H7.1) of hypothesis H7).

DEFINITION (under either (NH) or (H)). Define $\lambda_n = \inf\{\alpha \le \tau_D^{(n)}: S(n,\alpha) > 0$ or $\alpha = \tau_D^{(n)}\}$ (where $\tau_D^{(n)} = \inf\{\gamma: Z(n,\gamma) \in D\}$), $\zeta_n = S(n,\lambda_n)$, and $\zeta = \lim_n \zeta_n$.

Note (n,λ_n) is decreasing and so $\zeta_n \downarrow \zeta$. From Theorem 5.3(v) we have $\zeta = 0$ a.s. under condition (NH).

LEMMA 6.2. *Under condition* (NH), $\lim_n u(Z(n,\lambda_n)) = 0$ *a.s. on* $\{S(1,D) > 0\}$.

PROOF. This follows easily from the fact $\zeta_n - \zeta_{n+1} > 0$ for infinitely many n a.s. on $\{S(1,\tau_D) > 0\}$ (since $\zeta = 0$ a.s.) and the Markov property (4.2). See [4], Proposition 6.9, for details. □

PROOF OF THEOREM 6.1. Choosing increasing compact $C_n \subset \{y: u(y) < a/2\}$ and $\delta_n \downarrow 0$ such that as $n \to \infty$

(6.2) $$\sup\{e_{B(y,\delta_n)^c \cup \Delta}(y): y \in C_n\} \to 0$$

and

(6.3) $$P^x(S(1,\tau_D) > 0,\ Z(n,\lambda_n) \notin C_n) \downarrow 0.$$

These are possible by condition (NH) and Lemma 6.1 respectively. Let $D_n = C_n \cup D$ (note $C_n \subset D^c$). Since $P^x(\zeta_n) \downarrow 0$ and $S(n,\alpha)$ is increasing in (n,α) we have from (6.3) $P^x(S(n,\tau_{D_n}^{(n)})) \downarrow 0$; for $\tau_{D_n}^{(n)} \le \lambda_n$ if $Z(n,\lambda_n) \in D_n$. We proceed to produce a contradiction to H7.1) by proving

(6.4) $$H_{D_n}(x,D) > a/(2-a) = \varepsilon \quad \text{for all } n$$

and

(6.5) $$e_{D_n}(x) \downarrow 0.$$

Let $F = D_m$ for a fixed m. Consider the $\{F\}$-refinement $\tilde{Z}_\infty$ of Z_∞. We prove (6.5) first. Let

$$\hat{\tau}_F^{(n)} = \inf\{\gamma : \tilde{Z}(n,\gamma) \in F\},$$

$$\tilde{\tau}_F^{(n)} = \inf\{\tilde{\alpha} = \tilde{\alpha}(n) : Z(n,\alpha) \in F\}.$$

($\tilde{\alpha} = \tilde{\alpha}(n)$ was defined in the proof of Lemma 5.1); they will be denoted respectively by σ_1, σ_2 when $n = m$. Let $\tilde{R}(n,\alpha)$ be defined in $\tilde{Z}_\infty$ in the same way as $R(n,\alpha)$ in Z_∞. Then $P^x(\tilde{R}(m,\sigma_1)) \le P^x(\tilde{R}(m,\sigma_2))$ since $\sigma_1 \le \sigma_2$. Now $P^x(\tilde{R}(m,\sigma_1)) = e_F(x)$ by (4.4), and since $\tilde{Z}_m$ is a refinement of Z_m we have $P^x(\tilde{R}(m,\sigma_2)) = P^x(R(m,\tau_F^{(m)}))$. But the last expression equals $P^x(S(m,\tau_F^{(m)}))$, which then dominates $e_F(x)$. Since $P^x(S(n,\tau_{D_n}^{(n)}) \downarrow 0$, (6.5) follows. To prove (6.4) let m, F and $\tilde{Z}_\infty$ be as above; define a modification $\bar{S}$ of $S(1,\tau_D)$ as follows (all quantities in Z_∞ are regarded as defined in $\tilde{Z}_\infty$):

$$\bar{S} = \tilde{S}(n,\hat{\tau}_F^{(n)}) + S(1,\tau_D) \circ \tilde{\theta}(n,\hat{\tau}_F^{(n)}) \tag{6.6}$$

where the $\tilde{\theta}(n,\gamma)$ denote the shifts in $\tilde{Z}_\infty$ and $\tilde{S}(n,\gamma)$ are defined in $\tilde{Z}_\infty$ in the same way as $S(n,\gamma)$ in Z_∞ (note (6.6) is independent of n). We have

LEMMA 6.3. $\bar{S} \le S(1,\tau_D)$ *a.s.* P^x.

REMARK. Actually $\bar{S} = S(1,\tau_D)$ a.s. P^x since it is easy to see $P^x(\bar{S}) = e_D(x) = P^x(S(1,\tau_D))$.

PROOF. For $n \ge m$ let $\gamma_n = \inf\{\gamma \ge \hat{\tau}_F^{(n)} : \gamma = \tilde{\alpha}(n)$ for some $\alpha\}$. Let n_k be an increasing sequence of positive integers such that $R(n_k,n,\alpha)$ and $\tilde{R}(n_k,n,\gamma)$ converges a.s. P^x (to $S(n,\alpha)$ and $\tilde{S}(n,\gamma)$ respectively) for all (n,α) and (n,γ). Then

$$\bar{S} = \lim_k \tilde{R}(n_k,n,\hat{\tau}_F^{(n)}) + S(1,\tau_D) \circ \theta(n,\gamma_n) + d_n$$

where d_n is the difference between the second terms of the right-hand sides of (6.6) and the above display. Of course $d_n \ge 0$; from (6.2) one can show if $n > 1/\delta_N$ (and $n \ge m$), so that for $\hat{\tau}_F^{(n)} < \gamma < \gamma_n$ one has

$d(\tilde{Z}(n,\hat{\tau}_F^{(n)}),\tilde{Z}(n,\gamma)) < \delta_N$, and $N \geq m$, so that $C_m \subset C_N$, $P^x(d_n)$ is bounded by the supremum in (6.2) with n there replaced by N. Since d_n decreases a.s. P^x we have $d_n \downarrow 0$ a.s. P^x. By the definition of $R(n,\alpha)$ and $\tilde{R}(n,\gamma)$ we clearly have

$$\lim_k \tilde{R}(n_k,n,\hat{\tau}_F^{(n)}) \leq S(n,\alpha_n)$$

a.s. P^x where $\alpha_n = \inf\{\alpha : \tilde{\alpha}(n) \geq \hat{\tau}_F^{(n)}\}$, i.e. α_n is the ordinal α such that $\gamma_n = \tilde{\alpha}(n)$. It follows that $\bar{S} \leq S(1,\tau_D)$ a.s. P^x. □

To complete the proof of (6.4), recall $u(y) = P^y(S(1,\tau_D) = 0)$ and $u = 1$ on D. From Lemma 6.3 and the definition of $\bar{S}$ we thus have

$$u(x) \leq \int H_F(x,dy)u(y).$$

Since $F = D_m = C_m \cup D$ and m is arbitrary,

$$a = u(x) \leq \int H_{D_n}(x,dy)u(y)$$

$$\leq \frac{a}{2} H_{D_n}(x,C_n) + H_{D_n}(x,D) = \frac{a}{2} + (1 - \frac{a}{2})H_{D_n}(x,D).$$

Thus $H_{D_n}(x,D_n) \geq a/(2-a)$ for all n and (6.4) is proved. The proof of Theorem 6.1 is complete. □

We now prove the version of Theorem 6.1 under condition (H).

THEOREM 6.4. *Assume condition* (H). $P^x(S(m,1) > 0) = 1$ *for all* $x \neq \Delta$ *and* $m \geq 1$.

Again assume $m = 1$ and let $D = D(1,x)$; we show $u(x) = 0$ where u is as defined above. As in Theorem 6.1 we assume $u(x) = a > 0$ and show there exist increasing D_n satisfying (6.4) and (6.5). Here $P^x(\zeta > 0)$ is not necessarily true; without loss of generality assume it is positive. Let $Y = \lim_n Z(n,\lambda_n)$, which exists a.s. by Theorem 4.2 (note if $\lambda_n \in \mathcal{L}$, $Z(n,\lambda_n-) = \lim_{\alpha \uparrow \lambda_n} Z(n,\alpha)$). We will see below that a.s.

P^x on $\{\zeta > 0\}$. $Y \in H$ and $Y \neq \lim_n Z(n,\lambda_n)$ (so that (n,λ_n) is constant from some n on).

LEMMA 6.5. *Let* $\varepsilon > 0$. *There exists* $c > 0$ *and a subsequence* n_i *(which may be taken as a subsequence of* $n_{j(k)}(x)$ *in the definition of* $S(m,\beta)$) *and for each* i *a* Z_{n_i}*-stopping time* τ_i *such that* (n_i,τ_i) *increases,* $(n_i,\tau_i) \in \mathfrak{T}(n_{i-1},\lambda_{n_{i-1}})$ *and* $h(n_i,\tau_i,n_{i-1}) > c$ *a.s.* P^x *on* $\{\zeta > \varepsilon\}$ *except on a subset of which with* P^x*-measure less than* ε.

PROOF. We have $\lim_k R_k(m,\lambda_m-) = 0$ and $\lim_k R_k(m,\lambda_m) > \varepsilon$ a.s. P^x on $\{\zeta > \varepsilon\}$; now use the definition of $R_k(m,\beta)$ and apply an argument similar to (but simpler than) that in the proofs of Sublemmas 5.1.1 and 5.1.2. □

If $F_1 \in \mathfrak{D}$ with $F_1 \subset H^c$, using the lemma and an argument similar to that in the proof of Lemma 5.1 one can show $Y \in K - F_1$ a.s. P^x on $\{\zeta > 0\}$. It follows that $Y \in H$ a.s. P^x on $\{\zeta > 0\}$. For $c > 0$ let $\delta > 0$ be such that $P^x(\zeta;\Gamma) < c$ if $P^x(\Gamma) < \delta$ and $\Gamma \subset \{\zeta > 0\}$. Let $F \in \mathfrak{D}$ with $F \subset H \cup \Delta$ and $P^x(\zeta > 0,\ Y \notin F) < \delta$. Consider the $\{F\}$-refinement $\tilde{Z}_\infty$ of Z_∞. It is easy to establish $\tilde{S}(n,\gamma) \le S(n,\gamma)$ a.s. P^x for $\gamma \le \hat{\tau}^{(n)}_{F\cup D}$; thus $\tilde{S}(n,\hat{\tau}^{(n)}_{F\cup D} \wedge \lambda_n) \le S(n,\lambda_n) = \zeta_n$ a.s. P^x. Using condition H7.3) of hypothesis H7) one can show that a.s. P^x on $\{\zeta > 0,\ Y \in F\}$, $\tilde{S}(n,\hat{\tau}^{(n)}_{F\cup D} \wedge \tilde{\lambda}_n) \downarrow 0$. It follows that

$$\lim_n P^x(\tilde{S}(n,\hat{\tau}^{(n)}_{F\cup D} \wedge \tilde{\lambda}_n)) < c.$$

Considering the path behavior in $\tilde{Z}_\infty$, clearly $Y \neq \lim_n Z(n,\lambda_n)$ a.s. P^x on $\{\zeta > 0,\ Y \in F\}$. Thus a.s. P^x on $\{\zeta > 0,\ Y \in F\}$, (n,λ_n) must be constant from some n on; let N be the first such n. Repeat this argument with x replaced by $x_1 = Z(N,\lambda_N)$ if $u(x_1) \ge a/2$. Doing this a finite number of steps, and combining the argument used in the case of condition (NH) to treat the sets $\{\zeta = 0\}$, $\{\zeta > 0,\ \zeta\circ\theta(N,\lambda_N) = 0\}$, etc., we obtain

LEMMA 6.6. *There exist increasing compact* $F_m \subset H \cup \Delta$ *and* $C_m \subset H^c \cap \{u < a/2\}$ *such that* (i) $\inf\{h(y)\colon y \in F_m - \Delta\} = b_m > 0$, (ii) $\{C_m\}$ *satisfies* (6.2) *for some* $\delta_m \downarrow 0$, *and* (iii) *with* $D_m = F_m \cup C_m \cup D$ *we have* $e_{D_m}(x) \downarrow 0$.

To prove (6.4) for the sequence D_n defined above, we fix m and consider the $\{D_m\}$-refinement $\tilde{Z}_\infty$ of Z_∞.

LEMMA 6.7. $P^x(S(1,\tau_D) = 0,\ \hat{\tau}^{(n)}_{F_m} < \hat{\tau}^{(n)}_D) = 0$ *(note the event is independent of* n*).*

PROOF. Recall the sequences $m_{j(k)}(x)$ and $n_{j(k)}(x)$ in the definition of $S(m,\beta)$ which we denote by m_k, n_k respectively. Define $\beta_k = \inf\{\beta\colon \tilde{\beta}(m_k) \geq \hat{\tau}^{(m_k)}_{F_m \cup D}\}$. It is easy to show that $P^x(\bar{R}(n_k,m_k,\beta_k);\Gamma) \geq b_m P^x(\Gamma)$ for $\Gamma \subset \{\hat{\tau}^{(m_k)}_F \leq \hat{\tau}^{(m_k)}_D\}$, using (4.3) in $\tilde{Z}_\infty$. Since $\bar{R}(n_k,m_k,\hat{\tau}^{(m_k)}_D) \to 0$ a.s. P^x the lemma follows. □

To complete the proof of (6.4) we define as in (6.6)

$$\bar{S} = \tilde{S}(n,\hat{\tau}^{(n)}_{C \cup D}) + S(1,\tau_D) \circ \theta(n,\tau^{(n)}_{C \cup D})$$

in the $\{C \cup D\}$-refinement $\tilde{Z}_\infty$ of Z_∞ where $C = C_m$ for a fixed m. One again shows that Lemma 6.3 holds. For a fixed m let $F = F_m$, $C = C_m$. Let $\tilde{Z}_\infty$ be the $\{F \cup D,\ C \cup D\}$-refinement of $\tilde{Z}_\infty$. From Lemma 6.7 we have (in this refinement) $\hat{\tau}^{(n)}_{F \cup C \cup D} = \hat{\tau}^{(n)}_{C \cup D}$ (independent of n) a.s. P^x on $\{S(1,\tau_D) = 0\}$; therefore

$$\begin{aligned} u(x) &= P^x(S(1,\tau_D) = 0,\ \bar{S} = 0) \\ &= P^x(S(1,\tau_D) = 0,\ S(1,\tau_D) \circ \tilde{\theta}(n,\hat{\tau}^{(n)}_{C \cup D}) = 0) \\ &= P^x(S(1,\tau_D) = 0,\ S(1,\tau_D) \circ \tilde{\theta}(n,\hat{\tau}^{(n)}_{F \cup C \cup D}) = 0) \\ &\leq P^x(S(1,\tau_D) \circ \tilde{\theta}(n,\hat{\tau}^{(n)}_{F \cup C \cup D}) = 0) = \int H_{F \cup C \cup D}(x,dy)u(y). \end{aligned}$$

But $u = 0$ on F and $u < a/2$ on C; so as in the final computation in the proof of Theorem 6.1 we have $H_{F \cup C \cup D}(x,D) \geq a/(2-a)$, and (6.4) for the sets $D_m = F_m \cup C_m \cup D$ is proved.

7. The process (X_t)

We are now in a position to define a process (X_t), which under condition (NH) is the desired right process. Under condition (H) it has all the features of the desired process, except that all the holding times are in integrated form. In this case the desired right process can be defined on the product space of the sample space of (X_t) and countably many copies of $[0,\infty)$; such a construction is sketched in [5], section 5. This section is divided into three parts: first, we define (X_t); second, show that (X_t) has the correct hitting distributions; finally, prove (X_t) is strong Markov. From definition it will be clear that $E^x(T_\Delta) = e(x)$; therefore our construction will be complete. The proofs under condition (H) will only be sketched.

7.1 DEFINITION OF (X_t)

Assume condition (NH). Recall the mapping $(n,\alpha) \to Z_{n\alpha}$ on $\mathfrak{T}$ is right continuous (and has no oscillation), the mapping $(n,\alpha) \to S_{n\alpha}$ is strictly increasing, and its range $\mathcal{S}$ is dense in $[0,S_\Delta]$ where $S_\Delta = \sup \mathcal{S}$, all holding a.s. (again a.s. P^x for all x), (we assume holding surely). Therefore the following definition is justified.

DEFINITION. Assume condition (NH); define for all $t \geq 0$

$$\begin{aligned} X_t &= Z_{n\alpha} && \text{if } t = S_{n\alpha} \\ &= \lim_{S_{n\alpha} \downarrow t} Z_{n\alpha} && \text{if } t \in [0,S_\Delta] - \mathcal{S} \\ &= \Delta && \text{if } t > S_\Delta . \end{aligned}$$

Let $\mathcal{F} = \bigcap_\mu \sigma(Z_{n\alpha}, n \geq 1, \alpha < \pi)^\mu$, i.e. the intersection of all completions of $\sigma(Z_{n\alpha}, n \geq 1, \alpha < \pi)$ by the measures $P^\mu = \int \mu(dx) P^x$, μ a probability measure on $\mathcal{B}$. It is routine to show that the X_t are measurable relative to $\mathcal{B}^*$ and $\mathcal{F}$. Clearly the paths $t \to X_t$ are right continuous and have left limits.

Assume now condition (H). $\mathcal{S}$ is not necessarily dense in $[0, S_\Delta]$. We call a maximal open interval I contained in $[0, S_\Delta]$ with $I \cap \mathcal{S} = \emptyset$ an $\mathcal{S}$-free interval. List the (random) $\mathcal{S}$-free intervals as $I_1, I_2, \ldots$ by their lengths and ordering (in a measurable manner); if only $k-1$ such intervals exist, $I_k = I_{k+1} = \cdots = \emptyset$. Let $s_j = \inf I_j$, $r_j = \sup I_j$, with the convention $s_j = r_j = S_\Delta$ if $I_j = \emptyset$. Let $Y_j = \lim_{S_{n\alpha} \uparrow s_j} Z_{n\alpha}$ (which stands for $Z_{n\alpha}$ if $S_{n\alpha} = s_j$) and $W_j = \lim_{S_{n\alpha} \downarrow r_j} Z_{n\alpha}$. For $y \in H$, let $H(y, \cdot)$ denote the weak limit of $H_D(y, \cdot)$ as $D \uparrow K - \{y\}$ with $\overline{K - D} \downarrow \{y\}$ (which always exists); recall $h(y)$ is (meant to be) the expected holding time at y. The following definition is justified by Theorem 7.1 below.

DEFINITION. For each j define on $\{I_j \neq \emptyset\}$:

$$h_j = h_j(\omega) = \lim_m \lim_{S_{n\alpha} \uparrow s_j} e_{D(m, Y_j)}(Z_{n\alpha}),$$

$$\lambda_j(\cdot) = \lambda_j(\omega, \cdot) = \text{w-}\lim_m \text{w-}\lim_{S_{n\alpha} \uparrow s_j} H_{D(m, Y_j)}(Z_{n\alpha}, \cdot),$$

$$p_j = h_j / h(Y_j), \quad q_j = 1 - p_j.$$

Note that if $s_j = S_{n\alpha}$, then h_j and λ_j simply stand for $h(Y_j)$ and $H(Y_j, \cdot)$ respectively and $Y_j = Z_{n\alpha}$.

Fix P^μ. As in the treatment of the set $\{\zeta > 0\}$ in Theorem 6.4, we can define countably many sequences $\{(n_{ik}, \tau_{ik}), k \geq 1\}$ in $\mathfrak{T}$, $i = 1, 2, \ldots$, such that for each i, n_{ik} is strictly increasing in k, τ_{ik}

is a $Z_{n_{ik}}$-stopping time, (n_{ik},τ_{ik}) increases in k, $\lim_m \lim_k h(n_{ik},\tau_{ik},m) > 0$ on $\{\tau_{ik} < \pi$ for all $k\}$ and

$$\{\lim_k S(n_{ik},\tau_{ik}): i=1,2,\ldots\} = \{s_j: j=1,2,\ldots\}$$

a.s. P^μ. Therefore for each i there exists $j = j(i)$ such that $s_j = \lim_k S(n_{ik},\tau_{ik})$ and each j equals some $j(i)$. Let $\mathfrak{G}_i = \bigvee_k \mathfrak{G}(n_{ik},\tau_{ik})$; note this definition depends on the initial measure μ.

THEOREM 7.1. *Assume condition* (NH). *The following holds a.s. on* $\{I_j \neq \emptyset\}$: (i) $Y_j \in H$ *and* $r_j \in \mathfrak{S}$; (ii) h_j *exists,* $h_j = |I_j|$ *(the length of* I_j*) and* $0 < h_j \le h(Y_j)$ *(so that* $0 < p_j \le 1$*);* (iii) $\lambda_j(\cdot) \ge p_j H(Y_j,\cdot)$. *Moreover, for any* μ, $P^\mu(W_{j(i)} \in \cdot\,|\mathfrak{G}_i) = \lambda_{j(i)}(\cdot)$ *on* $\{I_{j(i)} \neq \emptyset\}$ *where* $\mathfrak{G}_i$ *and* $j(i)$ *are as defined above corresponding to* μ.

We omit the proof. $Y_j \in H$ a.s. on $\{I_j \neq \emptyset\}$ follows from a similar argument as in the proof of Theorem 6.1. For the rest let $F \in \mathbf{D}$ with $h > c$ for some $c < 0$ on $F - \Delta$ (in particular $F - \Delta \subset H$) and let $\tilde{Z}_\infty$ be the $\{F\}$-refinement of Z; then consider the successive hits of F in $\tilde{Z}_\infty$ and use H7.3) of H.7) and H4A.2) of H4A). The assertions for $Y_j \in F$ will follow. Incidentally, let $\nu_j(\cdot)$ be the measure satisfying

$$\nu_j(\cdot) = q_j^{-1}\,(\lambda_j(\cdot) - p_j H(Y_j,\cdot)),$$

which stands for the zero measure if $q_j = 0$ (in which case $\lambda_j = H(Y_j,\cdot)$); one can also establish that a.s. on $\{I_j \neq \emptyset,\ Y_j \in F\}$

$$\nu_j(\cdot) = \text{w-}\lim_m \ \text{w-}\lim_{S_{n\alpha}\uparrow S_j} H_{D(m,Y_j)\cap F}(Z_{n\alpha},\cdot).$$

DEFINITION. Assume condition (NH). By replacing Ω by $\Omega \times 2^\infty$ (2^∞ is the set of sequences of 0's and 1's) we may assume that Ω admits random variables δ_j assuming values 1 and 0 such that for each μ (with $\mathfrak{G}_i$ and $j(i)$ as defined above):

(i) $P^\mu(\delta_{j(i)} = 1|\mathfrak{G}_i) = p_{j(i)}$, $P^\mu(\delta_{j(i)} = 0|\mathfrak{G}_i) = q_{j(i)}$

(ii) $P^\mu(W_{j(i)} \in \cdot|\mathfrak{G}_i, \delta_{j(i)}) = \delta_{j(i)}H(Y_{j(i)},\cdot) + (1-\delta_{j(i)})\nu_{j(i)}(\cdot)$

(iii) $\delta_{j(i)}$ is conditionally independent of $(X_{r_j+t}, t \geq 0)$ given $W_{j(i)}$ under P^μ.

(These are possible because of Theorem 7.1.)

DEFINITION. Assume condition (NH). Define first

$$S'_{n\alpha} = S_{n\alpha} + \sum_j (\delta_j h(Y_j) - h_j) \cdot 1_{[s_j < S_{n\alpha}]},$$

$$\mathfrak{S}' = \{S'_{n\alpha}: n \geq 1, \alpha < \pi\}, \quad S'_\Delta = \sup \mathfrak{S}'.$$

Then define

$$\begin{aligned} X_t &= Z_{n\alpha} && \text{if } t = S'_{n\alpha} \\ &= \lim_{S'_{n\alpha}\downarrow t} Z_{n\alpha} && \text{if } t \notin \mathfrak{S}' \text{ but } (t,t+c)\cap\mathfrak{S}' \neq \emptyset \text{ for all } c > 0 \\ &= \lim_{S'_{n\alpha}\uparrow s} Z_{n\alpha} && \text{otherwise, where } s = \sup(\mathfrak{S}' \cap [0,t)). \end{aligned}$$

Let $\mathcal{F} = \bigcap_\mu \sigma(Z_{n\alpha}, n \geq 1, \alpha < \pi;\ \delta_j, j \geq 1)^\mu$. Then the X_t are measurable relative to $\mathcal{B}^*$ and $\mathcal{F}$. Clearly the paths $t \to X_t$ are right continuous and have left limits. Also, it is easy to see that $P^x(S'_{n\alpha}) = P^x(S_{n\alpha})$ for all x, n, α; in particular $P^x(S'_\Delta) = e(x)$.

7.2 HITTING DISTRIBUTION OF (X_t)

Recall the hitting time of D for (X_t) is

$$T_D = \inf\{t \geq 0: X_t \in D\}.$$

Note for $D \in \mathfrak{D}$, $T_D \leq S < \infty$ a.s. under condition (NH) and $T_D \leq S'_\Delta < \infty$ a.s. under (H).

THEOREM 7.2. *Assume condition* (NH). *For any* x *and* $D \in \mathfrak{D}$, $P^x(X(T_D) \in \cdot) = H_D(x,\cdot)$.

PROOF. Consider the $\{D\}$-refinement $\tilde{Z}_\infty$ of Z_∞. Of course $P^x(\tilde{Z}(n,\hat{\tau}_D^{(n)}) \in \cdot) = H_D(x,\cdot)$ (recall $\hat{\tau}_D^{(n)} = \inf\{\gamma: \tilde{Z}(n,\gamma) \in D\}$). Let $\tilde{T}_D = \inf\{S_{n\alpha}: \tilde{\alpha}(n) \geq \hat{\tau}_D^{(n)}\}$ (note $\tilde{S}_{n\alpha}$ is not used here). Since $\mathfrak{S}$ is dense in $[0,S_\Delta]$, the right continuity of $\tilde{Z}(n,\gamma)$ and that of (X_t) clearly imply $X(\tilde{T}_D) = \tilde{Z}(n,\hat{\tau}_D^{(n)})$ (the latter is of course independent of n). Therefore it suffices to show $\tilde{T}_D = T_D$ a.s. P^x. Of course $T_D \leq \tilde{T}_D$ a.s. Let $\Lambda = \{T_D < \tilde{T}_D\}$ and assume $P^x(\Lambda) > 0$. A.s. P^x on Λ there exist β_n such that (n,β_n) decreases and (n,β_n-) increases, $\tilde{\beta}_n(n) < \hat{\tau}_D^{(n)}$ (equivalently $S(n,\beta_n) < \tilde{T}_D$ a.s.) for sufficiently large n, and $\lim_n Z(n,\beta_n) = \lim_n Z(n,\beta_n-) \in D$. From this one can define (with a reasoning somewhat similar to the one in the proof of Theorem 3.4) disjoint compact $C_n \subset B(D,1/n) - D$ such that with $D_n = C_n \cup D$ and with any sequence of compact $F_m \uparrow K - D$ the conclusion of H4A.2) of hypothesis H4A) fails; we omit the details. □

The argument in the above proof clearly implies that under (NH) if $D \in \mathfrak{D}_n$ then $T_D = S(n,\tau_D^{(n)})$ a.s. so that $P^x(T_D) = e_D(x)$ (see Theorem 5.3); in particular $P^x(T_\Delta) = e(x)$.

THEOREM 7.3. *Assume condition* (H). *For any* x *and* $D \in \mathfrak{D}$, $P^x(X(T_D) \in \cdot) = H_D(x,\cdot)$.

PROOF. There exist increasing compact $C_n \subset D \cap H^c$ and $C_n' \subset D \cap \{h > 1/n\}$ (in particular $C_n' \subset H$) such that $T_{C_n \cup C_n' \cup \Delta} \downarrow T_D$ a.s. P^x and so $P^x(X(T_{C_n \cup C_n' \cup \Delta}) \in \cdot)$ converges weakly to $P^x(X(T_D) \in \cdot)$, and such that $H_D(x,C_n \cup C_n' \cup \Delta) \uparrow 1$ and so $H_{C_n \cup C_n' \cup \Delta}(x,\cdot)$ converges strongly (therefore weakly) to $H_D(x,\cdot)$ by H2). Therefore we may assume $D = D_1 \cup D_2$ where $D_1 \in \mathfrak{D}$ with $D_1 \subset K - H$ and $D_2 \subset \{h > c\}$ for some $c > 0$, D_2 compact. Consider the $\{D\}$-refinement $\tilde{Z}_\infty$ of Z_∞. By examining the behavior of $\tilde{Z}_\infty$ and (X_t) at the successive times when the paths contact or visit D_2, using the definition of (X_t) and Theorem 7.1 (see especially the remark after the theorem about how it can be proved), and applying the arguments in the proof of Theorem 7.2, one can argue that $P^x(X(T_D) \in \cdot) = P^x(\tilde{Z}(n,\hat{\tau}_D^{(n)}) \in \cdot) = H_D(x,\cdot)$. We omit the details. □

Again it is easy to see, using the fact $P^y(S'_{n\alpha}) = P^y(S_{n\alpha})$, that $P^x(T_D) = e_D(x)$ for $D \in \mathcal{D}_\infty$, in particular $P^x(T_\Delta) = e(x)$, under condition (H).

7.3 STRONG MARKOV PROPERTY OF (X_t)

Under either condition (NH) or (H), let $\mathcal{F}_t$ and $\mathcal{F}$ be defined from (X_t) in the usual way: first let $\mathcal{F}_t^0 = \sigma(X_s,\ s \le t)$, $\mathcal{F}^0 = \sigma(X_s,\ s < \infty)$; then define $\mathcal{F} = \bigcap_\mu (\mathcal{F}^0)^\mu$ where $(\mathcal{F}^0)^\mu$ is the completion of $\mathcal{F}^0$ under P^μ, and $\mathcal{F}_t = \bigcap_\mu (\mathcal{F}_t^0)^\mu$ where $(\mathcal{F}_t^0)^\mu$ denotes the σ-algebra generated by $\mathcal{F}_t^0$ and the P^μ-null sets of $(\mathcal{F}^0)^\mu$. It is easy to see that $\mathcal{F}$ defined above agrees with the $\mathcal{F}$ in §7.1 under either (NH) or (H). Let $\mathcal{F}_{t+} = \bigcap_{\varepsilon>0} \mathcal{F}_{t+\varepsilon}$. The strong Markov property of a right process (X_t) asserts that for all $(\mathcal{F}_{t+})$-stopping time T, $x \in K$, $\phi \in b\mathcal{F}$

$$P^x(\phi \circ \theta_T | \mathcal{F}(T)) = P^{X(T)}(\phi) \quad \text{on} \quad \{T < \infty\}. \tag{7.1}$$

We prove this under condition (NH), where (X_t) is a right process. Under condition (H) we prove that the above holds for all $(\mathcal{F}_{t+})$-stopping times T such that a.s. $T \notin I$ for any open interval I on which X_t is constant (and $X_t \in H$); this implies the strong Markov property of the right process defined from (X_t) of an enlarged sample space (see the beginning of the section).

THEOREM 7.4. *Assume condition* (NH). *The process* (X_t) *satisfies the strong Markov property.*

PROOF. It suffices to prove that for any x, $(\mathcal{F}^0_{t+})$-stopping time T, $f \in \mathbf{C}(K)$ with $f \geq 0$ and $t > 0$

$$P^x(f(X_{T+t}); T < \infty) = P^x(P^{X(T)}[f(X_t)], T < \infty).$$

We may assume $T \leq T_\Delta$ so that $T < \infty$ a.s. P^x. Since both sides of the above are right continuous in t it suffices to show that their Laplace transforms are equal. Thus denoting for $a > 0$

$$U^a f(y) = \int_0^\infty e^{-at} P^y[f(X_t)]dt = P^y \int_0^\infty e^{-at} f(X_t)dt$$

we show

$$P^x \int_0^\infty e^{-at} f(X_{T+t})dt = P^x(U^a f(X_T)). \tag{7.2}$$

Define $T_n = \inf\{S_{n\alpha} : S_{n\alpha} > T\}$ if $T < T_\Delta$ and $T_n = T_\Delta$ otherwise. Then $T_n > T$ if $T < T_\Delta$ and, since $\mathcal{S}$ is dense in $[0,T\ ]$ a.s., $T_n \downarrow T$ a.s. We claim that (7.2) holds with T replaced by T_n. This follows from the Markov property (4.2) with $\eta = \int_0^\infty e^{-at} f(X_t)dt$ and $\tau = \inf\{\alpha: S(n,\alpha) > T\}$ if $T < T_\Delta = S_\Delta$ and $\tau = \tau_\Delta^{(n)}$ otherwise. It is routine to show $\{\tau \leq \alpha\} \in \mathcal{G}(n,\alpha)$, so that τ is a stopping time relative to $(\mathcal{G}(n,\alpha), \alpha < \pi)$. Now it is clear that (7.2) will follow if one can show $U^a f(X_{T_n}) \to U^a f(X_T)$ a.s. P^x. Suppose the contrary; then for some $b_1 < b_2$ and compact $C \subset \{U^a f < b_1\}$ (resp. $C \subset \{U^a f > b_2\}$) we have, on a set Λ with $P^x(\Lambda) > 0$, $X_T \in C$ but $\overline{\lim_n}\, U^a f(X_{T_n}) > b_2$ (resp. $\underline{\lim_n}\, U^a f(X_{T_n}) < b_1$). Now from the Markov property (4.2) and a familiar supermartingale argument, $\lim_n U^a f(X_{T_n})$ exists a.s. P^x. For definiteness assume $C \subset \{U^a f < b_1\}$ and $\lim_n U^a f(X_{T_n}) > b_2$ on Λ. Furthermore for convenience assume as we may that there exist increasing compact $D_n \in \mathcal{D}$ with $D_n - \Delta \subset \{U^a f > b_2\}$ such that $X_{T_n} \in D_n - \Delta$ for all n on Λ. We will obtain a contradiction to hypothesis H5). To this end we establish two lemmas.

LEMMA 7.5. *Assume condition* (NH). *Let* $D \in \mathbf{D}$ *and* $S = T_D$. *Let* $\mathbf{G}(S)$ *be the* σ*-algebra generated by* X_S *and sets of the form* $\{Z_{m\beta} \in B, S_{m\beta} \le S\}$ ($\mathbf{G}(S)$ *is essential* $\sigma(X_{S\wedge t}, t < \infty)$). *Then for all* y, (i) $P^y(X_S \in \cdot) = H_D(y,\cdot)$ *(this restates Theorem* 7.2*)*; (ii) *we have the following Markov property at time* S: *for all* $\phi \in b\mathcal{F}$

$$P^y(\phi \circ \theta_S | \mathbf{G}(S)) = P^{X(S)}(\phi).$$

PROOF. To prove (ii) (again) consider the $\{D\}$-refinement $\tilde{Z}_\infty$ of $\tilde{Z}_\infty$. Using (i), the Markov property of Z_∞ at $\hat{\tau}_D$ and the argument of Lemma 6.3 (with D playing the role of F there), one can easily establish (ii).□

LEMMA 7.6. *Assume condition* (NH). *Let* $f \in b\mathbf{B}^*$ *and* $a > 0$. *Then* $U^a f(X(S_{n\alpha}))$ *converges to* $U^a f(y)$ *a.s.* P^y *as* $S_{n\alpha} \to 0$.

PROOF. As already observed $U^a f(X(S_{n\alpha}))$ converges a.s. P^y as $S_{n\alpha} \to 0$ (by the Markov property (4.2) and a supermartingale argument for $f \ge 0$). Let $T_n = T_{D(n,y)}$ which is S_{n1} a.s. P^y. We must show $\lim_n U^a f(X(T_n)) = U^a f(y)$ a.s. P^y. Now

$$U^a f(y) = \lim_m P^y[U^a f(X(T_m))] = \lim_m H_{D(m,y)} U^a f(y).$$

On the other hand it is easy to see that for $\varepsilon > 0$

$$\overline{\lim_n}\, P^y(|U^a f(X(T_n)) - H_{D(m,y)} U^a f(X(T_n))| > \varepsilon) < \varepsilon$$

if m is sufficiently large. Therefore it suffices to show for any fixed m

$$H_{D(m,y)} U^a f(X(T_n)) \to H_{D(m,y)} U^a f(y) \quad \text{a.s. } P^y.$$

But this follows from H5) (actually H5A)) with $D_n = D(n,y)$, $F = D(m,y)$, f there replaced by $U^a f$ and x by y, since $(X(T_n))$ is (equivalent to) the (W_n) there and $X(T_n) \to y$ a.s. P^y. □

To continue the proof of Theorem 7.4, let $\Lambda_1 = \Lambda \cap \{X_t \in K - C$ for all $t \in (T-\delta, T)$ for some $\delta > 0\}$ and $\Lambda_2 = \Lambda - \Lambda_1$. Then $P^x(\Lambda_1) > 0$ or $P^x(\Lambda_2) > 0$. Case 1). Assume $P^x(\Lambda_1) > 0$. Then there exists (n,α) such that $P^x(\Gamma) > 0$ where

$$\Gamma = \Lambda_1 \cap \{S_{n\alpha} < T,\ T = S_{n\alpha} + T_C \circ \theta_{S(n,\alpha)}\}.$$

Apply Markov property (4.2) at $S_{n\alpha}$; then apply Lemma 7.5 at $T_{C\cup\Delta}$ (i.e. at $T_{C\cup\Delta}\circ\theta_{S(n,\alpha)}$). Finally using Lemma 7.6 with $y = X(T_{C\cup\Delta})\circ\theta_{S(n,\alpha)}$ we obtain $U^a f(X(S_{n\alpha})) \to U^a f(X(T))$ as $S_{n\alpha} \downarrow T$ a.s. P^x on Γ. This contradicts $\Gamma \subset \Lambda_1 \subset \Lambda$. Case 2). Assume $P^x(\Lambda_2) > 0$. We claim that a.s. P^x on Λ_2, $X_t \in K - \bigcup_n D_n$ for all t in some interval $(T-\delta,\delta)$ ($\delta > 0$ depending on ω). For if not there must exist n_k such that with $V_k = \bigcup_{n=1}^{n_k} D_n$ and with $T_1, T_2, T_3, \ldots, T_{2k}, \ldots$ denoting successive hitting times of $C \cup \Delta$, V_1, $C \cup \Delta, \ldots, V_k, \ldots$ we have $T_k < T_{k+1}$ for all k on a subset of Λ_2 with positive P^x-measure which we assume to be Λ_2; but this is impossible since $U^a f(X(T_k))$ must converge a.s. P^x ($\{e^{-aT_k}U^a f(X(T_k))\}$ is a supermartingale by Lemma 7.5 — this is the supermartingale argument we have mentioned above, here in a simpler situation). By shifting x to an appropriate point $X(S_{n\alpha})$ we may assume $T_{D_n} > T$ for all n on Λ_2. Now apply hypothesis H5) to x and the sequence D_n, noting $(X(T_{D_n}))$ is (equivalent to) the (W_n) in H5) and $\lim_n X(T_{D_n}) = X(\lim_n T_{D_n})$ which is $X(T)$ on Λ_2. With $U^a f$ as the f in H5) and with F running through a sequence F_m such that a.s. P^x on Λ_2 it has a subsequence F_{m_j} satisfying $X(T) \in F_{m_j}$ and $\overline{K - F_{m_j}} \downarrow \{X(T)\}$, and using the same argument as in the proof of Lemma 7.6, we obtain $U^a f(X(T_{D_n})) \to U^a f(X(T))$ a.s. on Λ_2. But this contradicts $\Lambda_2 \subset \Lambda$. The proof of Theorem 7.4 is complete. □

THEOREM 7.7. *Assume condition* (H). *The process* (X_t) *is strong Markov in the sense that* (7.1) *is satisfied for all* x, $\phi \in b\mathcal{F}$ *and*

$(\mathcal{F}_{t+})$-stopping time T *such that* $T \notin I$ *for any open interval* I *on which* X_t *is constant.*

PROOF. Again one need only consider finite T. We may assume either $X(T) \in K - H$ a.s. P^x or $X(T) \in H \cup \Delta$ a.s. P^x. In the first case the proof is essentially the same as under condition (NH). In the second case we may assume further that $X(T) \in D$ a.s. P^x for some $D \in \mathcal{D}$ with $h > c$ on $D - \Delta$ for some $c > 0$. Then the strong Markov property as stated in the theorem follows from the Markov property as stated in Lemma 7.5 where S are successive hitting times of D ($S = S_n$ where $S_0 = 0$, $S_{n+1} = \inf\{t > S_n: X_t \in D\} \wedge T_\Delta$); this last fact can be proved together with Theorem 7.1 (see the remark following that theorem). We omit the details. □

8. Some pathological examples

Here we collect a few examples to show the nonvacuous necessity and mutual independence of the conditions of Theorems 1 and 2. In each example $\{H_D(x,\cdot): D \in \mathcal{D}, x \in K\}$ is the family of hitting distributions of a process (X_t) on a compact metric space K (containing a death point Δ) which may or may not be a legitimate right process. In some examples a function $e(x)$ is defined which is the expected lifetime function of the process or a modified version of it.

EXAMPLE 1. This example provides a family $\{H_D(x,\cdot)\}$ and a function e satisfying all conditions of Theorem 2 except H4A.2). First let $K_1 = [0,1] \cup \{2\}$ and consider a Markov process (X_t^1) on K_1 such that on $[0,1]$ it is Brownian motion reflected at 1 and absorbed at 0, and 2 is a holding point with mean holding time $h(2) = 1$ from which a jump is made to 0 w.pr. 1. Now delete the point 1 from K, and put 2 in its place; the process (X_t^1) is then denoted by (X_t), which has $K = [0,1]$ as its state space with $\Delta = 0$ and which is undefined at times t when $X_t^1 = 1$. Let $H_D(x,\cdot)$ be the hitting distributions and e the expected lifetime function of (X_t). All conditions of Theorem 2 are satisfied except H4A.2); for with $D = \{1,0\}$, $D_n = [1 - \frac{1}{n}, 1] \cup \{0\}$, $F_m = [0, 1 - \frac{1}{m}]$ we have ν_m concentrated on the point $(y,z) = (1, 1 - \frac{1}{m})$.

EXAMPLE 2. This and the next examples provide families $\{H_D(x,\cdot)\}$ satisfying all conditions of Theorem 1 except H5); this example does not even satisfy H5A) but the next does. Let $K = [-1,1]$ with $\Delta = 1$. Consider a process (X_t) on K such that on $(0,1]$ it is uniform motion to the right, on $[-1,0)$ it is uniform motion to the left, -1 is a holding point from which a particle jumps to Δ w.pr. 1, and 0 is an instantaneous point from which a particle moves to the right or to the left each w.pr. 1/2. The family $\{H_D(x,\cdot)\}$ of hitting distributions of (X_t) fails to satisfy H5A) with $x = 0$, $D_n = K - \left(-\frac{1}{n}, \frac{1}{n}\right)$, $F = \{-1,1\}$ and $f = 1_{(0,1]}$.

EXAMPLE 3. Let $K = \{(x_1,x_2)\colon 0 \le x_1,x_2 \le 1\}$ with $\Delta = (1,1)$. (X_t) is the process described as follows. A particle in $\{(x_1,x_2)\colon 0 < x_2 < 1\}$ moves upward with speed 1 to $(x_1,1)$; then moves to the right with speed 1 until reaching Δ. On $\{(x_1,0)\colon 0 \le x_1 \le 1\}$ a particle moves to the right until $S \wedge T_{(1,0)}$ where S is an independent exponential time and $T_{(1,0)}$ is the hitting time of the point $(1,0)$, at which time the particle moves directly upward. The family $\{(H_D(x,\cdot)\}$ of hitting distributions of (X_t) fails to satisfy H5) with $x = (0,0)$, $D_n = [0,1] \times [\frac{1}{n}, 1]$, $F = [0,1] \times \{1\}$, and $f = 1_{[0,\frac{1}{2}]\times\{1\}}$; but H5A) is satisfied.

EXAMPLE 4. In this example we have a family $\{H_D(x,\cdot)\}$ and a function e satisfying all conditions of Theorem 2 except H7.1) (at an instantaneous point x); but it does satisfy H7.1A). Let $K = \{(x_1, x_2): 0 \le x_1, x_2 \le 1\}$ with $\Delta = (1,1)$. (X_t) is the following process: each $x \in K - \Delta - \{(x_1,0): 0 \le x_1 < 1\}$ is a holding point with $h(x) = 1$ from which a jump is made to Δ w.pr. 1; on $\{(x_1,0): 0 \le x_1 < 1\}$ a particle moves to the right until the time $S \wedge T_{(1,0)}$ where S is an independent exponential time with mean 1, and when $S < T_{(1,0)}$ it jumps (immediately) to a point in K with uniform distribution on the square. Let $\{H_D(x,\cdot)\}$ be the family of hitting distributions of (X_t) and $e(x) \equiv 1$ for $x \neq \Delta$, $e(\Delta) = 0$ (note $e(x) = P^x(T_\Delta)$ if $x_2 > 0$ and $e(x) = P^x(T_\Delta - S \wedge T_{(1,0)})$ if $x_2 = 0$). H7.1) fails with $x = (0,0)$, $D = \{(x_1,x_2): x_1 = 1\}$, $D_n = \left\{(x_1,x_2): x_1 = 1 \text{ or } x_2 \ge \frac{1}{n}\right\}$ since $H_{D_n}(x,D_1) \ge H_D(x,(1,0)) = e^{-1}$ and $e_{D_n}(x) = (1 - e^{-1})/n$.

EXAMPLE 5. Let $K = [0,1]$ with $\Delta = 1$. Let $\{H_D(x,\cdot)\}$ be the hitting distributions of the process (X_t) on K which is uniform motion to the right until reaching Δ. Let $e(x) = 2$ for $x = 0$ and $= 1 - x$ for $0 < x \le 1$. Then $\{H_D(x,\cdot)\}$ and e satisfy all conditions of Theorem 2 except H7.2).

EXAMPLE 6. Here we have a family $\{H_D(x,\cdot)\}$ and a function e satisfying all conditions of Theorem 2 except H7.3). K, Δ, and $\{H_D(x,\cdot)\}$ are as in Example 5; but $e(x) = 2 - x$ for $0 \le x < 1$ and (of course) $e(1) = 0$.

EXAMPLE 7. Finally we show an example of $\{H_D(x,\cdot)\}$ and e satisfying all conditions of Theorem 2 except H7.4). Let $K = [0,1]$ with $\Delta = 1$. Consider the following process on K: each $x \neq \Delta$ is a holding point with $h(x) = 1$; from an irrational x a jump is made to Δ w.pr. 1, but from a rational r_n, where $\{r_1, r_2, \ldots\}$ is an enumeration of the rationals in $[0,1)$, a jump is made to r_{n+1} or Δ each w.pr. 1/2. Let

$\{H_D(x,\cdot)\}$ be its hitting distributions, and $e(x) = 1$ for x irrational and $e(r_n) = 2 + 2^n$ (note the expected lifetime of (X_t) is $P^x(T_\Delta) = 1$ for x irrational and $= 2$ for x rational but $\neq \Delta$). Then all conditions of Theorem 2 (including H7.3)) are satisfied, but H7.4) fails since with $x = r_1$, $D = \{r_n,\Delta\}$ we have $\int H_D(x,dy)e(y) = \frac{1}{2^{n-1}}(2^n + 2) > 2$.

References

1. R.M. BLUMENTHAL and R.K. GETOOR. *Markov Processes and Potential Theory.* Academic Press, New York, 1968.

2. R.K. GETOOR. *Markov Processes: Ray Processes and Right Processes.* Springer Lecture Notes in Math. No. 440. Springer, Berlin-Heidelberg-New York, 1975.

3. J.B. GRAVEREAUX and J. JACOD. Sur la construction des classes de processus de Markov invariantes par changement de temps, *Z. Wahrs. verw. Geb. 52* (1980), 75-107.

4. C.T. SHIH. Construction of Markov processes from hitting distributions. *Z. Wahrs. verw. Geb. 18* (1971), 47-72.

5. C.T. SHIH. Construction of Markov processes from hitting distributions II. *Ann. Math. Stat. 42* (1971), 97-114.

C. T. SHIH
Department of Mathematics
University of Michigan
Ann Arbor, MI 48109

Seminar on Stochastic Processes, 1983
Birkhäuser, Boston, 1984

REGULARITY PROPERTIES OF A STOCHASTIC PARTIAL DIFFERENTIAL EQUATION

by

JOHN B. WALSH

1. Introduction

This article is a sequel to [8] which studied the electrical potential in a randomly-stimulated neuron of the brain or spinal chord by means of the stochastic partial differential equation

$$\frac{\partial V}{\partial t} = \frac{\partial^2 V}{\partial x^2} - V + \dot{W}_{xt}, \tag{1.1}$$

where $\dot{W}$ is a space-time white noise. The neuron was represented by a line segment, and $V(x,t)$ was the electrical potential at x at time t.

Physically, the neuron is an object of finite diameter, and electrical conduction takes place at the membrane which forms its surface. It is logical to study the problem in two dimensions rather than one, and to consider, for example,

$$\frac{\partial V}{\partial t} = \frac{\partial^2 V}{\partial x^2} + \frac{\partial^2 V}{\partial y^2} - V + \dot{W}_{xyt}. \tag{1.2}$$

When this is done, one finds a quite different type of solution. The solution of (1.1) is a continuous function, but the solution of (1.2) is only a generalized function. This means that one must introduce the machinery of Schwartz distributions and Sobolev spaces to handle it.

The purpose of this article is to investigate the solutions of (1.2) and related equations, and in particular, to study their local smoothness properties. We will treat the problem in the following setting.

Let D be a bounded domain in $\mathbb{R}^n$ and let L be a uniformly elliptic self-adjoint differential operator with smooth coefficients on D. Let $\dot{W}_{xt}$ be a white noise on $D \times \mathbb{R}_+$ based on an underlying measure μ on D (see §2). Consider the parabolic stochastic partial differential equation,

$$\frac{\partial V}{\partial t} = LV + \dot{W}_{xt}, \quad x \in D,\ t > 0, \tag{1.3}$$

with zero initial condition and homogeneous linear boundary conditions on ∂D.

Kallianpur and Wolpert [7] have treated the case in which D is a differentiable manifold and, although they treat different questions than we do, this is perhaps the most logical setting. We have limited ourselves to middling generality for the sake of concreteness, but it is possible to extend our results to the manifold setting, for they are based on some elementary estimates on the Green's function (see §2) which are known to hold in manifolds [6].

The character of the solutions of (1.3) depends mainly on the dimension of D, and is relatively insensitive to perturbations of the operator L.

We will see that for each t, the solution $V(\cdot,t)$ of (1.3) can be regarded as a distribution on D. For an arbitrary underlying measure μ, V is the nth derivative of a continuous process: that is, there is a continuous function $u(x,t)$ such that $V = \frac{\partial^n}{\partial x_1 \cdots \partial x_n} u$ in the distributional sense. If more is known about μ, we can get more exact statements; if μ is a Lebesgue measure, for instance, V can be written as a derivative of order $n-1$ instead of n.

This implies that V_t takes its values in the Sobolev space $H_0^{-n+1}(D)$, but it says somewhat more. To say $V_t \in H^{-n+1}(D)$ tells one that V_t is the n-1st derivative of an L^2-function. However, we show that u is actually Hölder continuous, and we estimate its Hölder exponent.

The local behavior of V depends strongly on the local behavior of the underlying measure μ. In certain cases, the value $V(x,t)$ at a given point may exist. If $\mu(A) = 0$ for some open set A, then V is even smooth on A. If μ is Lebesgue measure, then $V(x,t)$ exists for all x if $n=1$; if $n = 2$ or 3, then $V(x,\cdot)$ exists as a distribution in t.

Apart from the use of distributions, our methods are much the same as in [8]. We have abandoned the eigenfunction expansion in favor of systematic exploitation of the Green's function, but this is just done to allow us to base our work on a minimal number of elementary facts about differential equations, collected for the most part in §2. We continue to use multiparameter stochastic integration as the main probabilistic tool. Smoothness results are usually obtained by estimating covariances and applying Gaussian process theory.

This contrasts with the usual method of attacking these problems by means of stochastic differential equations relative to Banach-valued Brownian motions ([7],[2]. See the bibliography of the latter for many further references). These approaches are more closely allied than it might appear. One requires an investment in functional analysis which it rewards with a simple and economical notation while the other needs only standard real variable theory but occasionally drags its user through thickets of iterated integrals. Still, the two approaches are nearly equivalent, at least in the present setting. Once posed, the problems can be solved either way. However, some problems are more natural to pose from one viewpoint than the other, and the most significant difference between the two may simply be that they lead to different questions.

The Banach-valued Brownian approach encourages one to regard the solution as a process $V(t)$ with values in a Banach space, and to ask functional-analytic questions: which space does $V(t)$ live in, and is it continuous there? Our real variable optic, on the other hand, leads us to attempt to peer past the rough exterior to see at the heart a function $V(x,t)$. It is then natural to ask questions about the joint behavior in the two variables and about the behavior in t for fixed x, as well as about the behavior in x for fixed t. These are largely the problems we treat here.

2. The Stochastic Integral Equation and the Green's Function

Let D be a relatively compact domain in $\mathbb{R}^n$ with a smooth boundary. Let L be a uniformly elliptic self-adjoint differential operator:

$$L = \sum_{i,j=1}^{n} a_{ij}(x)D_{ij} + \sum_{i=1}^{n} b_i(x)D_i + c(x), \qquad x \in D. \tag{2.1}$$

Let B be the operator

$$B = d(x)D_N + e(x), \qquad x \in \partial D, \tag{2.2}$$

where d and e are smooth functions and D_N is the normal derivative on the boundary.

If μ is a σ-finite measure on $\mathbb{R}^n$, let $\bar{\mu}$ be the measure $d\mu \times dt$ on $\mathbb{R}^{n+1}$. Let W be a Gaussian additive set function on the Borel sets of $\mathbb{R}^{n+1}$ such that $W(A)$ is a Gaussian random variable of mean zero and variance $\bar{\mu}(A)$, and such that if $A \cap B = \emptyset$, then $W(A)$ and $W(B)$ are independent. (We called W a white noise in [8]; we will reserve that term for something else here.) We will call μ the *underlying measure.*

Since D is bounded, we can and will assume, without loss of generality, that $D \subset \mathbb{R}^n_+$, the positive cone of $\mathbb{R}^n$. For $x = (x_1,\dots,x_n) \in \mathbb{R}^n_+$, define

$$W_{xt} = W\{(0,x_1]\times\cdots\times(0,x_n]\times(0,t]\}.$$

If μ is Lebesgue measure, then W has a version which is a continuous, real-valued, mean-zero Gaussian process called a *Brownian sheet.* Its covariance function is

$$E\{W_{xs}\, W_{yt}\} = (x_1 \wedge y_1)\cdots(x_n \wedge y_n)\cdot(s \wedge t).$$

In general, W is not continuous, but, as it is a strong martingale (in the multiparameter sense [9]), it will have a version which is right-continuous in (x,t).

Now W is not differentiable, but its derivative exists as a distribution. Let $\dot{W} = \dfrac{\partial^{n+1} W_{xt}}{\partial x_1 \cdots \partial x_n \partial t}$. We call $\dot{W}$ a *white noise.* One usually uses $\dot{W}$ in its integrated form; if $f(x,t)$ is a measurable, square integrable function,

$$\iint f(x,t)\dot{W}_{xt}\, dxdt = \iint f(x,t)dW_{xt},$$

where the latter is a stochastic Ito integral with respect to the Brownian sheet (see [1],[10], or the appendix of [8]). Since f is deterministic, we could take it to be either an Ito or Stratonovich integral, the two being identical in this case. For future reference, we recall the fact that

$$E\Big\{\Big(\iint f(x,t)dW_{xt}\Big)^2\Big\} = \int f^2(x,t)\mu(dx)dt$$

and, since f is deterministic, the stochastic integral is defined if and only if the latter integral is finite. Moreover, if f and g are square-integrable,

$$E\Big\{\Big(\int f(x,t)dW_{xt}\Big)\Big(\int g(x,t)dW_{xt}\Big)\Big\} = \int f(x,t)g(x,t)\mu(dx)dt.$$

We will use the following stochastic version of Fubini's theorem:

LEMMA 2.1. *If* $f(x,y,s)$ *is a measurable function on* $D\times D\times[0,\infty)$, *and if* ν *is a finite measure on* D *such that* $\int_0^\infty\int_D\int_D f^2(x,y,s)\nu(dx)\mu(dy)ds < \infty$, *then with probability one,*

$$\int_0^\infty\int_D\Big(\int_D f(x,y,s)\nu(dx)\Big)dW_{ys} = \int_D\Big(\int_0^\infty\int_D f(x,y,s)dW_{ys}\Big)\nu(dx).$$

(This is proved as usual, first showing that it holds for products of indicator functions, then for simple functions, and passing to the limit.)

We can now pose the following initial-boundary-value problem:

$$(2.3)\qquad \begin{aligned} &\frac{\partial V}{\partial t} = LV + \dot{W} && \text{in } D\times\mathbb{R}_+ \\ &BV = 0 && \text{on } \partial D\times\mathbb{R}_+ \\ &V(x,0) = 0 && \text{on } D \end{aligned}$$

This should be interpreted in a weak sense. Let us operate

purely formally for the moment. First integrate the PDE from 0 to t, then multiply by a test function $\phi(x)$ with $\operatorname{supp}(\phi) \subset D$, and integrate over x. If we define

$$V_t(\phi) = \int_D V(t,x)\phi(x)dx,$$

we get

$$V_t(\phi) - V_0(\phi) = \int_0^t (LV)_s(\phi)ds + \int_0^t \int_D \phi(x)\dot{W}_{xs}\, dxds.$$

This will make sense once we re-interpret the terms. Remembering that V is a distribution and that L is self-adjoint, we see $(LV)_s(\phi) = V_s(L\phi)$. Moreover, V satisfies the boundary conditions, so that this will hold for any test function ϕ which also satisfies the boundary conditions. The final integral is a stochastic integral as remarked above, so (2.3) becomes

$$(2.4) \qquad V_t(\phi) = \int_0^t V_s(L\phi)ds + \int_{D\times[0,t]} \phi(x)dW_{xs}, \quad t > 0$$

for all test functions ϕ on $\mathbb{R}^n$ which satisfy $B\phi(x) = 0$ for all $x \in \partial D$.

The condition that (2.4) hold for all ϕ satisfying $B\phi = 0$ is a translation of the boundary condition of (2.3), in the sense that if we replace $\dot{W}$ by a smooth function, so that V will be smooth, then (2.4) implies that $BV = 0$ on ∂D.

Let $G(x,t;y,s)$ be the Green's function of the homogeneous problem. That is, $G(x,t;y,s)$ is a positive function on $D\times\mathbb{R}_+ \times D\times\mathbb{R}$ such that G is continuous except at $(x,t) = (y,s)$, zero if $t < s$, and, as a function of (x,t) for fixed $(y,s) \in D\times\mathbb{R}_+$, G satisfies

$$\text{(2.5i)} \qquad \frac{\partial}{\partial t} G(\cdot,\cdot;y,s) = LG(\cdot,\cdot;y,s) \quad \text{on } D\times \mathbb{R}_+ - \{(y,s)\},$$

$$BG(\cdot,\cdot;y,s) = 0 \quad \text{on } \partial D\times \mathbb{R}_+, \qquad G(x,s;y,s) = \delta_y(x) .$$

We extend G to $\mathbb{R}_n \times \mathbb{R} \times \mathbb{R}_n \times \mathbb{R}$ by setting it equal to zero outside $\mathbb{D}\times\mathbb{R}_+\times\mathbb{D}\times\mathbb{R}$.

Let us introduce the notation

$$G(\phi,t;y,s) = \int \phi(x)G(x,t;y,s)ds.$$

Then we can rewrite (2.5i). If $B\phi = 0$ on ∂D,

$$\text{(2.5ii)} \qquad G(\phi,t;y,s) = \phi(y) + \int_s^t G(L\phi,u;y,s)du.$$

We recall some further facts about G [3].

$$\text{(2.6i)} \qquad G(x,t;y,s) = G(x,t-s,y,0)$$

$$\text{(2.6ii)} \qquad G(x,t;y,s) = G(y,t;x,s)$$

$$\text{(2.7)} \quad \int_D G(y,t;x,s)G(z,u;y,t)dy = G(z,u;x,s) \quad \text{if } s \le t \le u \text{ and } x,y,z \in D.$$

For $T > 0$ there exist positive constants C and δ such that [3], [6]:

$$\text{(2.8)} \qquad G(x,t;y,s) \le C(t-s)^{-\frac{n}{2}} e^{-\frac{|y-x|^2}{\delta(t-s)}}, \quad x,y \in D,\ s < t \le T;$$

$$\text{(2.9)} \quad \frac{\partial}{\partial x_1} G(x,t;y,s) \le C(t-s)^{-\frac{n+1}{2}} e^{-\frac{|y-x|^2}{\delta(t-s)}}, \quad x,y \in D,\ s < t \le T.$$

For any relatively compact set $K \subset D$ there exist strictly positive constants C and δ such that

$$\text{(2.10)} \qquad G(x,t;y,s) \ge C(t-s)^{-\frac{n}{2}} e^{-\frac{|y-x|^2}{\delta(t-s)}}, \quad x,y \in K,\ s < t \le T.$$

Moreover [3] if ϕ is bounded and continuous on $\mathbb{R}^n$, $G(\phi,t;y,s)$ is bounded and continuous on $y \in D$, $s < t < T$ for any T. This remains true if ϕ is only bounded and Borel measurable.

3. Existence, Uniqueness, and Smoothness in x

For a test function ϕ on $\mathbb{R}^n$, define

$$V_t(\phi) = \int_{D\times[0,t]} G(\phi,t;y,s)dW_{ys}. \tag{3.1}$$

This defines a mean zero Gaussian random variable of variance $\int_{D\times[0,t]} G^2(\phi,t;y,s)\mu(dy)ds$ and covariance function

$$E\{V_t(\phi)V_t(\psi)\} = \int_{D\times[0,t]} G(\phi,t;y,s)G(\psi,t;y,s)\mu(dy)ds \tag{3.1a}$$

where μ is the underlying measure. (The fact that the variance of $V_t(\phi)$ is finite follows from the facts that $G(\phi,t;y,s)$ is bounded and that $\mu(D) < \infty$.)

Now V_t is additive in ϕ, so by the Sazonov-Minlos theorem we can find a version of $\{V_t(\phi): \phi \in \mathcal{S}(\mathbb{R}^n)\}$ which is a distribution, i.e. for each u, $V_t(\cdot) \in \mathcal{S}'(\mathbb{R}^n)$. Let $V^{\mathcal{B}}$ be the restriction of V to the set $\mathcal{B} = \{\phi \in \mathcal{S}(\mathbb{R}^n): B\phi = 0 \text{ on } \partial D\}$.

THEOREM 3.1. $V^{\mathcal{B}}$ *satisfies* (2.4), *and* $V^{\mathcal{B}}$ *is uniquely determined by* (2.4).

PROOF. Suppose $B\phi = 0$ on ∂D. Note that

$$V_t(\phi) - \int_0^t V_s(\phi)ds - \int_{D\times[0,t]} \phi(y)dW_{ys}$$

$$= \int_{D\times[0,t]} [G(\phi,t;y,s) - \int_s^t G(L\phi;v;y,s)dv - \phi(y)]dW_{sy},$$

and the term in square brackets vanishes by (2.5ii).

The uniqueness proof is familiar, so we will cut corners slightly and use the (one-parameter) stochastic integral calculus. Suppose $\{U_t, t \geq 0\}$ satisfies (2.4) for all $\phi \in \mathbf{B}$. Note that $\{U_t(\phi), t \geq 0\}$ is a continuous semi-martingale, for $W_t(\phi) \overset{\text{def}}{=} \int_{D\times[0,t]} \phi(x)dW_{xt}$ is a continuous square-integrable martingale, and (2.4) can be written as

$$(3.2) \qquad dU_t(\phi) = U_t(L\phi)dt + dW_t(\phi).$$

Define $\psi_v(y) = G(\phi,t;y,v)$ and note that $\psi_v \in \mathbf{B}$ (to be more exact, ψ_v can be extended to $\mathbb{R}^n$ so that the extension is in $\mathbf{B}$) and $\frac{\partial\psi}{\partial v} = -L\psi$. Now $\phi \to U(\phi)$ is linear, so we can use Itô's formula to see that

$$d(U_s(\psi_s)) = (dU_s)(\psi_s) + U_s(\tfrac{\partial\psi}{\partial s}).$$

By (3.2) this is

$$\begin{aligned} &= U_s(L\psi_s)ds + dW_s(\psi_s) + U_s(\tfrac{\partial\phi}{\partial s})ds \\ &= U_s(-\tfrac{\partial\psi}{\partial s})ds + dW_s(\psi_s) + U_s(\tfrac{\partial\psi}{\partial s})ds \\ &= dW_s(\psi_s). \end{aligned}$$

Since $U_0(\psi) = 0$, this implies that

$$\begin{aligned} U_t(\phi) &= U_t(\psi_t) \\ &= \int_0^t dW_s(\psi_s) \\ &= \int_D \int_0^t G(\phi,t:y,s)dW_{ys} \\ &= V_t(\phi). \end{aligned}$$

Q.E.D.

We can now use (3.1) to extend the definition of V_t to functions other than test functions. Later we will define V on certain measures, but for the moment it is enough to note that $V_t(\phi)$ is defined by (3.1) for any bounded measurable function ϕ on $\mathbb{R}^n$.

One way of discussing the smoothness of a distribution T is to say that it lies in a given Sobolev space H^p.

If p is a positive integer, this means that all (weak) partials of T of order $\le p$ are L^2 functions, while if $p < 0$, it means that T is a sum of weak derivatives of order $\le p$ of L^2 functions. In our case, V is a continuous function if $n = 1$, but if $n \ge 2$ it lives in the Sobolev space H^{-n}. Even better, it is a (weak) derivative of order n of a Hölder continuous function.

Recall that a distribution T is the weak α^{th} derivative of a function $f \in L^2(\mathbb{R}^n)$ if for each test function ϕ,

$$T(\phi) = (-1)^{|\alpha|}\int f(x)D^{\alpha}\phi(x)dx.$$

PROPOSITION 3.2. *Suppose* $n \ge 2$. *Then there exists a process* $\{U(x,t),\ x \in D,\ t \ge 0\}$, *continuous in the pair* (x,t), *such that*

$$V(x,t) = \frac{\partial^n}{\partial x_1 \cdots \partial x_n} U(x,t)$$

in the weak sense.

PROOF. Define, for $x = (x_1,\dots,x_n) \in D$, $(0,x] = (0,x_1]\times\cdots\times(0,x_n]$ and set

$$K(x,y;s) = \int_{(0,x]} G(\xi,s;y,0)d\xi.$$

Then we define

$$U(x,t) = \int_0^t \int_D K(x,y;t-s)dW_{ys}.$$

We will show that U works.

In what follows, C is a strictly positive constant whose value may change from line to line.

First note that K is bounded on $\mathbb{R}_+^n \times (0,T]$ for each T, for it is dominated by $\int_{\mathbb{R}^n} G(\xi,s,y,0)d\xi$ which is bounded, by (2.8). Next,

$$|K(x,y,s) - K(x',y,s)| = \int_{(0,x]\Delta(0,x']} G(\xi,s,y,0)d\xi.$$

The symmetric difference $(0,x]\Delta(0,x']$ can be written as a union of n rectangles. If x_i and x_i' are the ith coordinates of x and x', respectively, and if $x_i < x_i'$, the ith such rectangle will be contained in $(x_i,x_i'] \times \mathbb{R}^{n-1}$. Thus by (2.8) the above is

$$\leq C \sum_{i=1}^{n} \Big|\frac{1}{\sqrt{s}} \int_{x_i}^{x_i'} e^{-\frac{(y_i-\xi_i)^2}{2\delta s}} d\xi_i\Big| \prod_{j\neq i} \frac{1}{\sqrt{s}} \int_{-\infty}^{\infty} e^{-\frac{|y_j-\xi_j|^2}{2\delta s}} d\xi_j$$

$$= C \sum_{i=1}^{n} \Big|\frac{1}{\sqrt{s}} \int_{x_i}^{x_i'} e^{-\frac{(y_i-\xi_i)^2}{2\delta s}} d\xi_i\Big|.$$

Set $\Delta = \frac{1}{2} \max|x_i' - x_i|$:

$$\leq C \frac{1}{\sqrt{s}} \int_0^{\Delta} e^{-r^2/2\delta s}\, dr.$$

Now

$$E\{(U(x,t) - U(x',t))^2\} = E\{(\int_D \int_0^t (K(x,y,t-s) - K(x',y,t-s))dW_{ys})^2\}$$

$$= \int_D \int_0^t (K(x,y,t-s) - K(x',y,t-s))^2 d\mu(y)ds$$

$$= \mu(D) \int_0^t |K(x',y,s) - K(x,y,s)|^2 ds$$

$$\leq C \int_0^t \frac{ds}{s}\Big(\int_0^{\Delta} e^{-r^2/2\delta s}\, dr\Big)^2.$$

We can bound the inner integral by integrating over a circle of radius

$\sqrt{2}\,\Delta$ in polar coordinates:

$$\le C \int_0^t (1 - e^{-\Delta^2/\delta s})ds$$

$$\le C \int_0^t 1 \wedge \frac{\Delta^2}{\delta s}\, ds$$

$$= C \int_0^{\Delta^2/\delta} ds + C \int_{\Delta^2/\delta}^{t} \frac{\Delta^2}{\delta s}\, ds$$

$$= C(\Delta^2 + \Delta^2(\log t - \log \Delta^2/\delta))$$

$$\le C\,\Delta^2 \log \frac{1}{\Delta} \qquad \text{for } \Delta << 1.$$

Next

$$E\{(U(x,t+s) - U(x,t))^2\} = \int_0^t \int_D (K(x,y;t+s-u) - K(x,y;t-u))^2 \mu(dy)du$$
$$+ \int_t^{t+s} \int_D K(x,y,t+s-u)^2\, \mu(dy)du.$$

Using similar methods — which we will omit here since we will be doing a similar but more delicate calculation in Theorem 3.5 — we can see that each of these integrals is bounded by a constant times s, hence this is

$$\le Cs$$

if s and t are bounded by T.

Thus, if $x,y \in D$ and $s \le T$, $t \le T$,

$$E\{(U(x,t) - U(y,s))^2\}^{\frac{1}{2}} \le C\left(|x-y|\sqrt{\log \frac{1}{|x-y|}} + \sqrt{|t-s|}\right).$$

We can now apply a theorem of Garsia, Rodemich, and Rumsey [4] to conclude that $U(x,t)$ has a version which is uniformly continuous in (x,t) on $D \times [0,T]$ for each T.

Now let us show that $V = \frac{\partial^n}{\partial x_1 \cdots \partial x_n} U$. We must show that $V_t(\phi) = (-1)^n\, U_t\left(\frac{\partial^n \phi}{\partial x_1 \cdots \partial x_n}\right)$ for each ϕ of compact support in D.

Since U is uniformly continuous on D,

$$U_t(\phi) = \int_D \phi(x)U(x,t)dx = \int_D \phi(x) \int_0^t \int_D K(x,y;t-s)dW_{ys}dx.$$

By Lemma 2.1 we can change the order of the stochastic and the Lebesgue integral:

$$= \int_0^t \int_D \Big[\int_D \phi(x)K(x,y,t-s)dx\Big]dW_{ys}$$

$$= \int_0^t \int_D \phi(x) \int_{[0,x)} G(\xi,t;y,s)d\xi \, dW.$$

Now integrate by parts over each of the n coordinates of x to see that this is

$$= \int_0^t \int_D (-1)^n \frac{\partial^n \phi(x)}{\partial x_1 \cdots \partial x_n} G(x,t;y,s)dx \, dW_{ys}$$

$$= (-1)^n V_t\Big(\frac{\partial^n \phi}{\partial x_1 \cdots \partial x_n}\Big) .$$

Q.E.D.

Proposition 3.2 can be sharpened in several ways: one can use the theorem in [4] to show that U is Hölder continuous, for instance, and one can deduce that V_t lives in the Sobolev space $H_0^{-n}(D)$, and that, moreover, $t \to V_t$ is continuous there. However, we shall leave this aside for now. We shall prove the analogous facts in a more interesting and delicate situation (see Theorem 3.5 and Corollary 3.8).

In fact, in most cases, V is far smoother than Proposition 3.2 would indicate. This is not surprising, since there are no restrictions on the underlying measure μ in Proposition 3.2, and it is μ which controls the noise.

We shall see what happens locally when we restrict μ in the following two propositions, and then we shall see what happens globally when μ is absolutely continuous with respect to Lebesgue measure in Theorem 3.5 and Corollary 3.8.

Let us first look at a single point x. We say V has *point values at* x if the stochastic integral $\int_D \int_0^t G(x,t;y,s)dW_{ys}$ converges for one (and therefore all) $t > 0$. This happens if and only if

$$(3.3) \qquad \int_D \int_0^t G(x,t;y,s)^2 ds\, \mu(dy) < \infty.$$

For a fixed x, let $\nu_x(r) = \mu\{y:|y-x| \le r\}$, $r \ge 0$.

PROPOSITION 3.3. *A necessary and sufficient condition that* V *have point values at* x *is that* $\nu_x(0) = 0$ *and*

$$\text{(i)} \quad \int_{0+} r^{2-2n}\, d\nu_x(r) < \infty \qquad \textit{if} \quad n \ge 2;$$

or

$$\text{(ii)} \quad \int_{0+} \log 1/r\, d\nu_x(r) < \infty \qquad \textit{if} \quad n = 1.$$

PROOF. If $x,y \in D$ and $t < T$,

$$G(x,t;y,s) \le C_1 t^{-\frac{n}{2}} e^{-\frac{|y-x|^2}{2\delta_1 t}}$$

while for y in a small ε-neighborhood of x and for $t < T$,

$$G(x,t;y,s) \ge C\, t^{-\frac{n}{2}} e^{-\frac{|y-x|^2}{2\delta_2 t}} .$$

Set

$$\begin{aligned} I(\varepsilon,\delta,x) &= \int_{|y-x|<\varepsilon} \int_0^1 (t-s)^{-n} e^{-\frac{|y-x|^2}{\delta(t-s)}} ds\, \mu(dy) \\ &= \int_0^\varepsilon \int_0^1 s^{-n} e^{-\frac{r^2}{\delta s}} ds\, d\nu_x(r). \end{aligned}$$

Change variables in the inner integral

$$= \delta^{n-1} \int_0^\varepsilon \int_{r^2/\delta}^\infty u^{n-2} e^{-u}\, du\, r^{2-2n}\, d\nu_x(r).$$

If $n \ge 2$, this is

$$= (n-2)!\,\delta^{n-1} \int_0^{\varepsilon} \frac{d\nu_x(r)}{r^{2n-2}} - \delta^{n-1} \int_0^{\varepsilon} \int_0^{r^2/\delta} u^{n-2} e^{-u}\, du\, \frac{d\nu_x(r)}{r^{2n-2}} .$$

Thus

$$(n-2)!\,\delta^{n-1} \int_0^{\varepsilon} r^{2-2n}\, d\nu_x(r) - \frac{\nu_x(\varepsilon)}{n-1} \le I(\varepsilon,\delta,x)$$

$$\le (n-2)!\,\delta^{n-1} \int_0^{\varepsilon} r^{2-2n}\, d\nu_x(r).$$

Thus $I(\varepsilon,\delta,x)$ is finite iff $\int_{0+} \frac{d\nu_x(r)}{r^{2n-2}} < \infty$; this is independent of $\varepsilon > 0$ and $\delta > 0$. If $n = 1$, a similar argument shows it is finite iff the integral in (ii) is finite.

But (3.3) is finite iff $I(\varepsilon,\delta,x) < \infty$, since it is bounded below by $C_2 I(\varepsilon,\delta_1,x)$ and above by $C_1 I(\varepsilon,\delta_1,x) + C_1 \int_{\varepsilon}^{\infty} \int_0^1 s^{-n} e^{-r^2/\delta_1}\, ds\, d\nu_x(r)$, and this latter integral is finite. Indeed, the above calculations show that it is smaller than $(n-2)!\,\delta^{n-1} \frac{1}{\varepsilon^{2n-2}} \mu(0) < \infty$. *Q.E.D.*

If it happens that μ does not charge some open set S, then V has point values everywhere in S. It is easy to see that V is even L^2 bounded in any compact subset K of S. Assuming we have picked a measurable version of V, Fubini's theorem assures us that V is actually an L^2 function on K. Suppose ϕ has compact support in S. Then (2.4) reduces to

$$V_t(\phi) = \int_0^t V_s(L\phi)\,ds. \tag{3.4}$$

If $\psi(x,s)$ is a C^∞ function of compact support in $S \times (0,\infty)$, set $V(\psi) = \int_S \int_0^\infty V(x,t)\psi(x,t)\,dt\,dx$. Noting that $V_t(\phi) = \int_D V(x,t)\phi(x)\,dx$, we can apply (3.4) to $-\frac{\partial\psi}{\partial t}$ to see that $V\left(\frac{\partial\psi}{\partial t} + L\psi\right) = 0$. Since $-\frac{\partial}{\partial t} + L$ is hypoelliptic, Hörmander's theorem implies that $V(x,t)$ is a C^∞ function. More exactly we have proved:

PROPOSITION 3.4. *If* S *is an open subset of* D *such that* $\mu(S) = 0$, *then* V *has point values in* S, *and* V *has a version such that* $\{V(x,t),\ x \in S,\ t > 0\}$ *is a.s. a* C^∞ *function.*

The most interesting case by far occurs when μ is Lebesgue measure, or more generally, when μ is absolutely continuous with respect to Lebesgue measure. If $n = 1$, V has point values, for then $\nu_x(r) = 2r$ and the integral in Proposition 3.3 converges. If $n = 2$, $\nu_x(r) = \pi r^2$ and the integral in Proposition 3.3 diverges, as it does for $n \geq 2$. The following theorem, the central result of this section, tells us how close V is to being a function: it is the $n-1^{st}$ mixed partial of a Hölder continuous function.

THEOREM 3.5. *Suppose* $n \geq 2$, *and suppose that* μ *is absolutely continuous with respect to Lebesgue measure with bounded density* $\sigma^2(x)$. *Then there exists a process* $\{U(x,t): x \in D,\ t \geq 0\}$ *which is a.s. Hölder continuous in* (x,t) *with exponent* $\frac{1}{4} - \varepsilon$ *for any* $\varepsilon > 0$, *such that*

$$V(x,t) = \frac{\partial^{n-1}}{\partial x_2 \cdots \partial x_n} U(x,t)$$

in the weak sense.

Before proving this we need some notation and a lemma. If $x = (x_1, \ldots, x_n)$ we will write $\bar{x} = (x_2, \ldots, x_n)$ for the last $n-1$ coordinates of x, and write $x = x_1\bar{x}$. We will let $(0,x]$ denote the n-dimensional rectangle $(0,x_1] \times \cdots \times (0,x_n]$, and, similarly, $(0,\bar{x}] = (0,x_2] \times \cdots \times (0,x_n]$. Recall that $D \subset \mathbb{R}^n_+$ so that if $x \in D$, all coordinates of x are positive.

LEMMA 3.6. *Let* $J = (a_2,b_2] \times \cdots \times (a_n,b_n]$ *and* $K = (\alpha_2,\beta_2] \times \cdots \times (\alpha_n,\beta_n]$ *be rectangles in* $\mathbb{R}^{n-1}$. *Let* $T_0 > 0$. *Then there are constants*

C and $\delta > 0$ *such that if* $0 \le s < t \le T_0$ *and* $\xi, \eta \in \mathbb{R}$,

$$\int_{J\times K} G(\xi\bar{u},r;\eta\bar{v},0)d\bar{u}d\bar{v} \le \frac{C}{\sqrt{r}} e^{-\frac{(\xi-\eta)^2}{\delta r}} \prod_{j=2}^{n} (b_j - a_j) \wedge (\beta_j - \alpha_j).$$

The same inequality holds if $G(x,t;y,0)$ *is replaced by*

$$g(x,y;t) \overset{\text{def}}{=} t^{-\frac{1}{2}} e^{-\frac{|y-x|^2}{\delta t}}.$$

PROOF. By (2.8) the integral is

$$\le \frac{C}{\sqrt{r}} e^{-\frac{(\xi-\eta)^2}{\delta r}} \prod_{j=2}^{n} \int_{a_j}^{b_j} \int_{\alpha_j}^{\beta_j} \frac{1}{\sqrt{r}} e^{-\frac{(u_j-v_j)^2}{\delta r}} du_j dv_j.$$

But now

$$\int_{\alpha_j}^{\beta_j} \frac{1}{\sqrt{r}} e^{-\frac{(u_j-v_j)^2}{\delta r}} du_j \le \int_{-\infty}^{\infty} \frac{1}{\sqrt{r}} e^{-\frac{(u_j-v_j)^2}{\delta r}} = \sqrt{2\pi\delta},$$

so the double integral is bounded by $\sqrt{2\pi\delta}(b_j - a_j)$. By symmetry, it is also bounded by $\sqrt{2\pi\delta}(\beta_j - \alpha_j)$ and the lemma follows.

PROOF (of Theorem 3.5). Write $x = \xi\bar{x}$ and define

$$H(x,y;s) = \int_{(0,\bar{x}]} G(\xi\bar{u},s;y,0)d\bar{u}.$$

Then put

$$U(x,t) = \int_0^t \int_D H(x,z,t-s)dW_{zs}.$$

We will show that $V = \frac{\partial^{n-1}}{\partial x_2 \cdots \partial x_n} U$. We will calculate three expectations:

$$(3.5)\qquad E\{U(x,t)^2\} = \int_0^t \int_D H(x,z,s)^2\, \sigma^2(z)dzds;$$

$$(3.6)\qquad E\{(U(y,t)-U(x,t))^2\} = \int_0^t \int_D (H(y,z,s)-H(x,z,s))^2\, \sigma^2(z)dzds;$$

$$(3.7)\qquad E\{(U(x,t)-U(x,s))^2\} = \int_0^s \int_D (H(y,z,t-r)-H(y,z,s-r))^2\, \sigma^2(z)dzdr + \int_s^t \int_D H(y,z,t-r)^2\, \sigma^2(z)dzdr,$$

if $s < t$. Here we have used the fact that W is independent on the disjoint sets $D\times[0,s]$ and $D\times(s,t]$. These expectations all depend on integrals of H, so the following lemma will be useful.

LEMMA 3.7. *Let* $\Gamma(x,t;y,s) = \int_0^{s\wedge t} \int_D H(x,z;t-r)H(y,z,s-r)dzdr$. *Then, if* $x = \xi\bar{x}$ *and* $y = \eta\bar{x}$,

$$(3.8)\qquad \Gamma(x,t;y,s) = \int_0^{s\wedge t} \int_{(0,\bar{x}]\times(0,\bar{y}]} G(\xi\bar{u},s+t-2r;\eta\bar{v},0)d\bar{v}\, d\bar{u}\, dr.$$

PROOF. From the definition of H,

$$\Gamma(x,t;y,s) = \int_0^{s\wedge t}\int_D \int_{(0,\bar{x}]\times(0,\bar{y}]} G(\xi\bar{u};t-r,z,0)G(\eta\bar{v};s-r,z,0)d\bar{v}\, d\bar{u}\, dzdr.$$

Integrate first over z and use (2.7) to get (3.8). *Q.E.D.*

We return to the proof of Theorem 3.2. We first note that, since σ^2 is bounded, (3.5) is bounded above by $\|\sigma^2\|_\infty\, \Gamma(x,t;x,t)$. Now apply Lemma 3.3 to the right hand side of (3.8) to see that

$$\Gamma(x,t;x,t) \le \int_0^{s\wedge t} \frac{C}{\sqrt{r}}\, dr < \infty.$$

Thus $U(x,t)$ exists. It is evidently a mean-zero Gaussian process.

In order to find its continuity properties, we turn to (3.6) and (3.7). From (3.6)

$$E\{(U(y,t)-U(x,t))^2\} \le \|\sigma^2\| \{[\Gamma(y,t;y,t)-\Gamma(y,t;x,t)]$$
$$+ [\Gamma(x,t;x,t) - \Gamma(x,t;y,t)]\}.$$

The two terms in square brackets are similar, so we will only consider the first one. Its absolute value is

$$\Big|\int_0^t \Big[\int_{(0,\bar y]\times(0,\bar y]} G(\eta\bar u, 2t-2r; \eta\bar v, 0)\,d\bar u\,d\bar v$$
$$- \int_{(0,\bar x]\times(0,\bar x]} G(\eta\bar u, 2t-2r, \xi\bar v, 0)\,d\bar u\,d\bar v\Big]dr\Big|$$

where we write $x = \xi\bar x$ and $y = \xi\bar y$. This is

$$\le \frac{1}{2}\int_0^{2t}\int_{(0,\bar x]\times(0,\bar x]} |G(\eta\bar u, r; \eta\bar v, 0) - G(\eta\bar u, r; \xi\bar v, 0)|\,d\bar u\,d\bar v\,dr$$
$$+ \frac{1}{2}\int_0^{2t}\int_{(0,\bar x]\times(0,\bar x]\Delta(0,\bar y]\times(0,\bar y]} G(\eta\bar u, r; \eta\bar v, 0)\,dr = J_1 + J_2$$

where "Δ" means the symmetric difference. Write

$$G(\eta\bar u, r; \eta\bar v, 0) - G(\eta\bar u, r; \eta\bar v, 0) = \int_\xi^\eta \frac{\partial G}{\partial \zeta}(\eta\bar v, r; \zeta\bar v, 0)\,d\zeta.$$

By (2.9)

$$\Big|\frac{\partial G}{\partial \xi}\Big| \le \frac{C}{\sqrt{r}}\, g(\eta\bar u, \zeta\bar v; r),$$

where g is defined in Lemma 3.3. Then

$$J_1 \le C\int_0^{2t}\frac{dr}{\sqrt{r}}\int_{(0,\bar x]\times(0,\bar x]} \Big|\int_\xi^\eta g(\eta\bar u, \xi\bar v; r)\,d\bar u\,d\bar v\Big|.$$

By Lemma 3.6, this is

$$\leq C \int_0^{2t} \frac{dr}{r}\Big(\prod_{j=2}^{n} x_j\Big)\int_\xi^\eta e^{-\frac{(\xi-\eta)^2}{2\delta r}}\, d\zeta.$$

Absorb the x_j in the constant and change variables:

$$= C|\xi - \eta| \int_0^{\sqrt{2\delta t}/|\xi-\eta|} du \int_0^{1/u} e^{-\zeta^2}\, d\zeta.$$

Now change order and estimate the resulting integrals:

$$J_1 \leq C\sqrt{t}\ |\xi - \eta|(1 + \log^+(1/|\xi - \eta|)).$$

Turning to J_2, let $\beta = \max\{x_j, y_j : j = 2,\ldots,n\}$ and note that $(0,\bar{x}]\times(0,\bar{x}]\Delta(0,\bar{y}]\times(0,\bar{y}]$ is contained in a union of sets of the form $K_k = (\alpha_4,\beta_4]\times\cdots\times(\alpha_{2n},\beta_{2n}]$, where $(\alpha_j,\beta_j] = (0,\beta]$ if $j \neq k$, and if $j = k = 2i$ or $2i+1$, then $(\alpha_k,\beta_k] = (x_i,y_i]$ (or $(y_i,x_i]$ if $x_i > y_i$).

By Lemma 3.6

$$\int_0^{2t}\int_{K_k} G(\eta\bar{u},r;\eta\bar{v},0)\,d\bar{u}\,d\bar{v}\,dr \leq C|x_i - y_i|\beta^{n-2}\int_0^{2t}\frac{dr}{\sqrt{r}},$$

hence

$$J_2 \leq C\sqrt{t}\sum_{j=2}^{n}|x_j - y_j|.$$

Thus, absorbing $\|\sigma^2\|$ in the constant, we see that

$$E\{(U(x,t) - U(y,t))^2\} \leq C\Big[|y_1 - x_1|(1+ |\log^+|y_1 - x_1||) + \sum_{j=2}^{n}|y_j-x_j|\Big] \leq C|y-x|(1 + \log^+\frac{1}{|y-x|}). \tag{3.9}$$

From (3.7) we see that if $s < t$,

$$E\{(U(x,t) - U(x,s))^2\} \leq \|\sigma^2\|[\Gamma(x,t;x,t) - 2\Gamma(x,t;x,s) + \Gamma(x,s;x,s)]$$

$$= \|\sigma^2\| \int_s^t \int_{(0,\bar{x}]\times(0,\bar{x}]} G(\xi\bar{u},2t-2r,\xi\bar{v},0)d\bar{u}\,d\bar{v}\,dr$$

$$+ \|\sigma^2\| \int_0^s \int_{(0,\bar{x}]\times(0,\bar{x}]} \Big[G(\xi\bar{u},2t-2r;\xi\bar{v},0) - 2G(\xi\bar{u},s+t-2r;\xi\bar{v},0)$$

$$+ G(\xi\bar{u},2s-2r;\xi\bar{v},0)\Big]d\bar{u}\,d\bar{v}\,dr \;=\; I_1 + I_2.$$

Change variables in I_1, letting $q = 2t-2r$, and use Lemma 3.6 to see that

$$I_1 \le C \int_0^{t-s} \Big(\prod_{j=2}^{n} x_j\Big)\frac{dq}{\sqrt{q}} \le C\sqrt{t-s}.$$

We now make the same type of change of variables in each of the terms of I_2. Integrate first over q. This will give us a difference of integrals of $G(\xi\bar{u},q;\xi\bar{v},0)$ over various intervals. There is cancellation, and we are left with

$$I_2 = 2\int_{(0,\bar{x}]\times(0,\bar{x}]} \Big[\int_0^{t-s} - \int_{t-s}^{2t-2s} + \int_{t+s}^{2t} - \int_{2s}^{t+s} G(\xi\bar{u},q;\xi\bar{v},0)dq\Big]d\bar{u}\,d\bar{v}.$$

Each of the integrals in brackets is over an interval of length $t-s$. By Lemma 3.6, each of these is dominated by $C\sqrt{t-s}\,\prod_{i=2}^{n} x_i$. Adding these together, and combining the result with I_1, we see that for $s < t$ and $x \in D$,

$$\text{(3.10)} \qquad E\{(U(x,t) - U(x,s))^2\} \le C\sqrt{t-s}.$$

Combining (3.9) and (3.10), we see that if $x,y \in D$ and $s,t \le T_0$, then

$$\text{(3.11)} \quad E\{(U(y,t)-U(x,s))^2\}^{\frac{1}{2}} \le C\Big[\sqrt{|y-x|\Big(1+\log^+\frac{1}{|y-x|}\Big)} + |t-s|^{\frac{1}{4}}\Big]$$

which is certainly

$$\le C|(y,t) - (x,s)|^{\frac{1}{4}}.$$

By [4] Theorem 2, $\{U(x,t),\ t > 0,\ x \in D\}$ has a continuous version, and there exists a constant d and a random variable X such that for all $x,y \in D$ and $s,t \le T_0$, if $\delta = |(y,t) - (x,s)|$

$$|U(x,t) - U(y,s)| \le X\delta^{\frac{1}{4}} + d\delta^{\frac{1}{4}}\sqrt{\log 1/\delta}. \tag{3.12}$$

It remains to verify that if $D^\alpha = \dfrac{\partial^{n-1}}{\partial x_2 \cdots \partial x_n}$ that $V = D^\alpha U$. To see this, we must show that $V(\phi) = (-1)^{n-1}U(D^\alpha\phi)$ for all test functions ϕ with compact support in D.

Since U is uniformly continuous on D,

$$U_t(\phi) = \int_D \phi(x)U(x,t)dx = \int_D \phi(x) \int_0^t \int_D H(x,y;t-s)dW_{ys}.$$

We can interchange the order

$$= \int_0^t \int_D \int_D \phi(x)H(x,y;t-s)dxdW_{ys}$$

$$= \int_0^t \int_D \phi(x_1,\ldots,x_n)\int_0^{x_2}\cdots\int_0^{x_n} G(x_1,u_2,\ldots,u_n;t;y,s)du_2\cdots du_n]dW_{ys}.$$

Integrate by parts over $x_2,\ldots,x_n$ successively to see that this is

$$= \int_0^t \int_D (-1)^{n-1}D^\alpha\phi(x)G(x,t;y,s)dxdW_{ys} = (-1)^{n-1}V_t(D^\alpha\phi).$$ *Q.E.D.*

COROLLARY 3.8. *Under the hypotheses of Theorem* 3.5 V_t *is a continuous process whose state space is the Sobolev space* $H_0^{1-n}(D)$.

PROOF. There is very little to prove once we recall the relevant definitions. $H_0^p(D)$ is the space of functions f of compact support in D whose partials of order $\le p$ are in L^2, and it has the norm

$$\| f \|_p = \sum_{|\alpha| \le p} \| D^\alpha f \|_{L^2}.$$

H_0^{-p} is the dual of $H_0^p(D)$ with the dual norm $\| \ \|_{-p}$. If $g \in C(D)$ and $T = D^\alpha g$, where $|\alpha| \le p$, then $T \in H_0^{-p}(D)$ and $\| T \|_{-p} \le \sqrt{|D|}\, \| g \|_\infty$, since for $f \in H_0^p(D)$,

$$|T(f)| = \langle g, D^\alpha f\rangle \le \| g\|_{L^2} \| D^\alpha f\|_{L^2} \le \sqrt{|D|}\ \| g\|_\infty \| f\|_p$$

where $|D|$ is the Lebesgue measure of D.

Now, since $V_t = D^\alpha U_t$ where $|\alpha| = n-1$, $V_t \in H_0^{1-n}(D)$ and

$$\| V_t - V_s\|_{1-n} \le \sqrt{|D|}\ \| U_t - U_s\|_\infty .$$

Since U is uniformly continuous in x and t by (3.12) we conclude that $t \to V_t$ is continuous. *Q.E.D.*

REMARKS. Theorem 3.5 gives a slightly misleading picture of the smoothness of U as a function of x for fixed t. If we use (3.9), then Theorem 2 of [4] tells us that for each fixed t there is a constant d and a random variable X such that for $x,y \in D$

$$|U(y,t) - U(x,t)| \le X\sqrt{|y-x|\left(1 + \log^+ \frac{1}{|y-x|}\right)} + d\sqrt{y-x}\ \log \frac{1}{|y-x|} .$$

Thus U is (nearly) Hölder 1/2, rather than Hölder 1/4 as the Theorem states.

The reason for the discrepancy is that the theorem deals with continuity in x *and* t, and $U(x,t)$ is only Hölder 1/4 as a function of t for fixed x.

The modulus of continuity of U is exactly that found for V in the one-dimensional case [8]. This is not surprising, since we have essentially integrated $n-1$ of the variables out of V to get U.

4. More on Point Values

We saw in §3 that V can have point values in certain cases. We now ask when $V(x,\cdot)$ can exist, not as a function, but as a distribution. We are going to specialize to what we think is the most interesting case: when the measure μ underlying the white noise $\dot{W}$ is Lebesgue measure. This will allow us to get some explicit formulas.

One way of phrasing this problem is to regard V as a distribution on $D \times \mathbb{R}_+$, rather than on D, and ask when it has a trace on the line x = constant.

Since we have a specific representation (2.4) of V, we can use this to bypass Sobolev theory and define the trace directly.

Let ν be a measure on $D \times \mathbb{R}_+$, let $D_\infty = D \times \mathbb{R}_+$, and set

$$G(\nu;y,s) = \int_{D_\infty} G(x,t;y,s)\nu(dxdt).$$

Let $\mathbf{E}^+$ be the class of measures ν for which $\int_{D_\infty} G^2(\nu;y,s)dyds < \infty$, and let $\mathbf{E} = \mathbf{E}^+ - \mathbf{E}^+$ be the class of differences of measures in $\mathbf{E}^+$. For $\nu \in \mathbf{E}$, define

$$V(\nu) = \int_{D_\infty} G(\nu;y,s)dW_{ys}. \tag{4.1}$$

Then $V(\nu)$ is a mean zero Gaussian random variable with variance

$$E\{V(\nu)^2\} = \int_{D_\infty} G^2(\nu;y,s)dyds;$$

one can see that this is finite by writing $\nu = \nu^+ - \nu^-$, where ν^+ and ν^- are in $\mathbf{E}^+$, and noting that $G(\nu;y,s) = G(\nu^+;y,s) - G(\nu^-;y,s)$. If ϕ is a test function on D and if $\nu(dxdt) = (\phi(x)dx)\delta_s(dt)$ (where δ_s is the unit mass at s) then $V(\nu) = V_s(\phi)$, so that this is an extension of V_t.

If ν_1, $\nu_2 \in \mathbf{E}$, then, remembering that $\mu(dx) = dx$,

$$E\{V(\nu_1)V(\nu_2)\} = \int_{D_\infty} G(\nu_1;y,s)G(\nu_2;y,s)dyds$$

$$= \int_{D_\infty}\Big[\int_{D_\infty}\int_{D_\infty} \nu_1(dxdt)G(x,t;y,s)G(x',t';y,s)dxdtdx'dt'\Big]dyds.$$

Integrate first over y and use (2.7). Remember that $G(x,t;y,s) = 0$ if $s > t$, so that this is

$$(4.2)\quad E\{V(\nu_1)V(\nu_2)\}$$
$$= \int_0^\infty\Big[\int_{D_\infty\times D_\infty} \nu_1(dxdt)G(x,t+t';x',2s)I_{\{t\wedge t'\geq s\}}\,\nu_2(dx'dt')\Big]ds.$$

We want to consider the behavior of $V(x,\cdot)$ for fixed x, so we consider measures ν of the form $\delta_x \times \phi dt$, where ϕ is a measurable function on $\mathbb{R}_+$, and define (when it makes sense)

$$V(x,\phi) = V(\delta_x \times \phi dt).$$

If $\delta_x \times \phi$ and $\delta_x \times \psi$ are in $\mathbf{E}$, $V(x,\phi)$ and $V(x,\psi)$ are mean zero Gaussian random variables with covariance

$$(4.3)\quad \Gamma(\phi,\psi) = E\{V(x,\phi)V(x,\psi)\}$$
$$= \int_0^\infty\int_0^\infty\int_0^\infty G(x;u+v;x,0)\phi(u+s)\psi(v+s)du\,dv\,ds.$$

Moreover, we can see that $\delta_x \times \phi dt \in \mathbf{E}$ iff

$$\int_0^\infty\int_0^\infty\int_0^\infty G(x,u+v;x,0)\phi(u+s)\phi(v+s)du\,dv\,ds < \infty.$$

Suppose ϕ is bounded and has support in $[0,T]$. By (2.8) there is a constant C, possibly depending on T, such that $G(x,t;x,0) \leq Ct^{-n/2}$, $x \in D$, $t \leq T$. Thus the above integral is

$$\leq C\|\phi\|_\infty^2 \int_0^T\int_0^{T-s}\int_0^{T-s} \frac{du\,dv\,ds}{(u+v)^{n/2}}\,.$$

This is finite if $n \le 3$ and infinite if $n \ge 4$. It follows that if $n \le 3$, $V(x,\phi)$ is defined for all test functions. By the Sazonov-Minlos theorem [5], we have

THEOREM 4.1. *If* $n \le 3$, *then for each* $x \in D$, $V(x,\cdot)$ *is a distribution. Its covariance function* $\Gamma(\phi,\psi)$ *is given by* (4.2).

5. Spectral Properties of $V(x,\cdot)$

We assume in this section that the underlying measure μ of the white noise is Lebesgue measure, and that the dimension of D is $n \le 3$. Then $V(x,\cdot)$ exists, either as a function ($n = 1$) or as a distribution ($n = 2$ or 3).

We would like to use the spectral theory of stationary processes on $V(x,\cdot)$. Now V is not stationary as is, but under mild conditions we can define the closely-related process $\tilde{V}(x,\cdot)$ by

$$\tilde{V}(x,\phi) = \int_D \int_{-\infty}^{\infty} G(\phi,t-s,y,0)\,dW_{ys}. \tag{5.1}$$

From (4.3) $\tilde{V}$ has covariance $\tilde{\Gamma}$ given by

$$\begin{aligned} \tilde{\Gamma}(\phi,\psi) &= E\{\tilde{V}(x,\phi)\,\overline{\tilde{V}(x,\psi)}\} \\ &= \int_{-\infty}^{\infty}\int_0^{\infty}\int_0^{\infty} G(x,u+v,x,0)\phi(u+s)\bar{\psi}(v+s)\,du\,dv\,ds. \end{aligned} \tag{5.2}$$

where we allow ϕ and ψ to be complex.

In order that (5.1) make sense for ϕ and ψ bounded and of compact support, we need some conditions on G. The following is sufficient:

$$\int_1^{\infty} tG(x,t;x,0)\,dt \quad \text{is bounded for } x \in D. \tag{5.3}$$

It follows from this that L has a potential kernel K:

(5.4) $$K(x,y) = \int_0^\infty G(x,t;y,0)dt \qquad x \neq y.$$

For example, if the functions c, d, and e in the expressions (2.1) and (2.2) satisfy $c(x) \le -\varepsilon < 0$, $e(x) > 0$ and $d(x) \ge 0$, for all x, then the operator $\hat{L} = L + \varepsilon$ will satisfy the maximum principle [3], hence if $\hat{G}$ is the Green's function corresponding to $\hat{L}$, then for all $t > 1$, $x,y \in D$, $\hat{G}(x,t;y,0) \le \sup_{x,y \in D} \hat{G}(x,1;y,0)$, and this is finite by (2.8). Since $G(x,t;y,s) = e^{-\varepsilon(t-s)}\,\hat{G}(x,t;y,s)$, (5.3) clearly holds.

Notice that under (5.3), V_t tends in distribution to $\tilde{V}_t$ as $t \to \infty$. To see this note that both are mean zero Gaussian processes, and that one can rewrite the covariance function of V_t given in (3.1a) just as we did in (4.2):

$$E\{V_t(\phi)V_t(\psi)\} = \frac{1}{2}\int_D\int_D\int_0^{2t} \phi(x)\psi(y)G(x,u;y,0)du\,dx\,dy.$$

Then (5.3) allows us to let $t \to \infty$. This converges to

$$\frac{1}{2}\int_D\int_D \phi(x)\psi(y)K(x,y)dx\,dy = E\{\tilde{V}_t(\phi)\tilde{V}_t(\psi)\}.$$

Thus we can expect V and $\tilde{V}$ to be similar processes, at least for $t > 0$. In particular, they will have the same local behavior, which is what we wish to study.

The reason for using $\tilde{V}$ rather than V is simply that we want to avail ourselves of the theory of stationary processes, and, in particular, of their spectral representation.

By [5] there exists a measure σ on $\mathbb{R}$, called the *spectral measure*, and a complex Gaussian random measure Z_λ on $\mathbb{R}$ such that $\tilde{V}$ has the spectral representation

(5.5) $$\tilde{V}(x,\cdot) = \frac{1}{\sqrt{2\pi}}\int_{-\infty}^{\infty} e^{-it\lambda}\,dZ_\lambda$$

in the sense that if ϕ is a test function and $\hat{\phi}$ its Fourier transform,

$$\tilde{V}(x,\phi) = \int_{-\infty}^{\infty} \hat{\phi}(\lambda)dZ_\lambda . \tag{5.6}$$

Here, if A is an interval, $Z(A)$ is a complex-valued Gaussian random variable with $E\{|Z(A)|^2\} = \sigma(A)$, and if A and B are disjoint intervals, $Z(A)$ and $Z(B)$ are independent. (Thus $\{Z_\lambda(0,t],\ t \geq 0\}$ is a Gaussian martingale and (5.5) is defined as a stochastic integral.)

THEOREM 5.1. *Suppose* $n \leq 3$. *Then the measure* σ *has a continuous density, denoted* $|p(\lambda)|^2$, *called the spectral density of* $\tilde{V}$, *and*

$$|p(\lambda)|^2 = \frac{2}{\lambda}\int_0^{\infty} G(x,u;x,0)\sin \lambda u\, du, \qquad \lambda > 0. \tag{5.7}$$

PROOF. Consider the function

$$\phi(t) = \frac{1}{\sqrt{2\pi}}\,\frac{e^{ibt} - e^{iat}}{t} .$$

Then $\hat{\phi}(\lambda) = I_{(a,b)}(\lambda)$ except at $\lambda = a$ and $\lambda = b$, so that, choosing a and b such that $\sigma\{a\} = \sigma\{b\} = 0$,

$$\sigma\{(a,b]\} = E\{(Z_b - Z_a)^2\} = E\Big\{\Big(\int \hat{\phi}\,dZ_\lambda\Big)^2\Big\} = E\{|\tilde{V}(x,\phi)|^2\}$$

$$= \int_0^{\infty}\int_0^{\infty} G(x,u+v;x,0)\Big(\int_{-\infty}^{\infty} \phi(u+s)\bar{\phi}(v+s)ds\Big)du\, dv.$$

The Fourier transform gives an L^2-isometry and $\widehat{\phi(u+\cdot)} = e^{i\lambda u}\hat{\phi}(\cdot)$ so

$$\int_{-\infty}^{\infty} \phi(u+s)\bar{\phi}(v+s)ds = \int_{-\infty}^{\infty} e^{i\lambda(u-v)}|\hat{\phi}(\lambda)|^2 d\lambda = \frac{e^{ib(u-v)} - e^{ia(u-v)}}{i(u-v)}$$

since $\hat{\phi} = I_{ab}$.

Split the integral over u and v into an integral over $u \leq v$ and another over $v < u$ and add. We get

$$(5.8) \quad \sigma\{(a,b]\} = 2\int_0^\infty du \int_u^\infty dv G(x,u+v,x,0)\left[\frac{\sin b(v-u) - \sin a(v-u)}{v-u}\right].$$

Now fix λ and choose $a < \lambda < b$ in (5.8). Divide by $b-a$, let $b \downarrow \lambda$, $a \uparrow \lambda$, and note that

$$\frac{\sin b(v-u) - \sin a(v-u)}{(v-u)(b-a)}$$

tends boundedly to $\cos \lambda(v-u)$. The integral converges absolutely by (5.3), so by dominated convergence we can go to the limit. We see

$$|p(\lambda)|^2 = \lim_{\substack{b\downarrow\lambda \\ a\uparrow\lambda}} \frac{\sigma\{(a,b]\}}{b-a} = 2\int_0^\infty du \int_u^\infty dv\, G(x,u+v,x,0)\cos \lambda(v-u)$$

$$= 2\int_0^\infty \cos \lambda u\left(\int_u^\infty G(x,u+v;x,0)\cos \lambda v\, dv\right)du$$

$$+ 2\int_0^\infty \sin \lambda u\left(\int_u^\infty G(x,u+v;x,0)\sin \lambda v\, dv\right)du.$$

Integrate by parts with respect to u in both integrals. After cancellation, this is

$$= \frac{2}{\lambda}\int_0^\infty G(x,u;x,0)\sin \lambda u\, du.$$

This holds for all λ, showing that the density exists and is given by (5.7).

To see that it is continuous in λ, decompose G as follows:

$$G(x,t;x,0) = ct^{-n/2}\, g(t) + h(t)$$

where g is differentiable, $g(0) = 1$ and $g(t) = 0$ for $t > 1$, while $h(t)$ is differentiable, $th(t)$ is integrable on $(0,a)$, and $h(0) = 0$. This can be done since $G(x,t;x,0) \sim ct^{-n/2}$ for small t, and G is differentiable in t and integrable on $[1,\infty)$ by (5.3). Then

$$(5.9) \qquad |p(\lambda)|^2 = 2c\int_0^1 u^{1-n/2}\, g(u)\, \frac{\sin \lambda u}{\lambda u}\, du + 2\int_0^\infty uh(u)\, \frac{\sin \lambda u}{\lambda u}\, du.$$

Since $1 - \frac{n}{2} \geq -\frac{1}{2}$, we can apply the dominated convergence theorem to both integrals to see that $|p(\lambda)|^2$ is continuous on $(0,\infty)$. *Q.E.D.*

COROLLARY 5.2. *For* $n \leq 3$, $|p(\lambda)|^2 \sim \dfrac{c}{\lambda^{2-n/2}}$ *as* $\lambda \to \infty$ *and* $\lim_{\lambda\to 0} |p(\lambda)|^2 = 2\int_0^\infty uG(x,u;x,0)du < \infty$.

PROOF. Write (5.9) as $|p(\lambda)|^2 = I(\lambda) + J(\lambda)$. Now as $\lambda \to \infty$, $J(\lambda) = O(\lambda^{-2})$ as one can see by integrating by parts, while if we set $q(\lambda) = \int_0^t u^{-n/2} \sin u\, du$ we can rewrite $I(\lambda)$ in the form

$$-\frac{2c}{\lambda^{2-n/2}} \int_0^1 g'(u)\, q(\lambda u)du.$$

As $\lambda \to \infty$, $q(\lambda u)$ tends boundedly to $q(\infty) = \int_0^\infty u^{-n/2} \sin u\, du$, which exists as a conditionally convergent (n = 1 or 2) or absolutely convergent (n = 3) integral. Thus $\lambda^{2-n/2}\, I(\lambda) \to -2cq(\infty)\int_0^1 g'(s)ds = 2cq(\infty)$.

This proves the first statement of the corollary. The second follows by letting $\lambda \to 0$ in (5.9) and using dominated convergence.

The behavior of the spectral density at ∞ determines the local smoothness of $\tilde{V}(x,\cdot)$. If n = 1, $\tilde{V}(x,\cdot)$ is Hölder continuous [8]. If n = 2 or 3, $\tilde{V}(x,\cdot)$ exists only as a distribution. We shall see that it is the weak derivative of a Hölder continuous function.

THEOREM 5.3. *Let* $x \in D$. *There exists a continuous process* $\{U(t), t \geq 0\}$ *such that on* $(0,\infty)$, $\tilde{V}(x,\cdot)$ *is the weak derivative of* U. *Moreover, there is a constant* C *and a random variable* X *such that* $\{U_t, 0 \leq t \leq 1\}$ *has a modulus of continuity dominated by* $m_2(\delta) =$

$C\delta \log 1/\delta + X\delta\sqrt{\log 1/\delta}$, $0 \le \delta \le 1$ *if* $n = 2$; *or* $m_3(\delta) = C\delta^{3/4}\sqrt{\log 1/\delta} + X\delta^{3/4}$, $0 \le \delta \le 1$ *if* $n = 3$.

PROOF. Define $U(t) = \tilde{V}(x, I_{[0,t]})$. This is a Gaussian process with stationary increments and

$$E\{(U(s+t) - U(s))^2\} = E\{U(t)^2\}$$

$$= E\{|\int_{-\infty}^{\infty}(\hat{I}_{[0,t]}(\lambda))dZ_\lambda|^2\}$$

where $\hat{I}_{[0,t]}(\lambda) = \frac{i}{\sqrt{2\pi}}\frac{e^{-it\lambda} - 1}{\lambda}$. This is

$$= \frac{1}{2\pi}\int_{-\infty}|\frac{e^{-it\lambda} - 1}{\lambda}|^2|p(\lambda)|^2 d\lambda$$

$$= \frac{t}{2\pi}\int_{-\infty}^{\infty}|\frac{e^{-iu} - 1}{u}|^2|p(\tfrac{N}{t})|^2 \, du.$$

Now $|\frac{e^{-iu} - 1}{u}| \le 1$. If $\varepsilon > 0$, we can find N large enough so that

$$\frac{c - \varepsilon}{\lambda^{2 - n/2}} \le |p(\lambda)|^2 \le \frac{c}{\lambda^{2 - n/2}} \qquad \text{for } |\lambda| \ge N$$

and p is bounded, by M, say, on $[-N,N]$. Thus the above is

$$\le Mt\int_{-Nt}^{Nt} du + \frac{ct^{3-n/2}}{\pi}\int_{Nt}^{\infty} u^{n/2-2}\, du.$$

If $t < 1/N$, this is

$$\le c_1t^2 + c_2t^{3-n/2}\int_{Nt}^{1} u^{n/2-2}\, du + c_3t^{3-n/2}\int_{1}^{\infty} u^{n/2-4}\, du.$$

If $n = 2$, this is

$$\le c_2t^2 \log 1/t + O(t^2)$$

and if $n = 3$, it is

$$\leq ct^{3/2} + O(t^2).$$

Now apply [4] to see that U has a continuous version with the given moduli of continuity.

To see that $\tilde{V}(x,\cdot)$ is indeed the weak derivative of U, let ϕ be a test function on $[0,\infty)$. Then

$$U_t = \tilde{V}(x,I_{[0,t]}) = \int_D \int_{-\infty}^{\infty} \Big(\int_0^t G(x,u;y,s)du \Big) dW_{ys},$$

so

$$\int_0^\infty U_t \, \phi'(t)dt = \int_0^\infty \int_D \int_{-\infty}^{\infty} \Big(\int_0^t G(x,u;y,s)du \Big) dW_{ys} \, \phi'(t)dt.$$

We can interchange order, then integrate by parts:

$$= \int_D \int_{-\infty}^{\infty} \int_0^\infty \phi'(t) \int_0^t G(x,u;y,s)du \, dW_{ys}$$

$$= -\int_D \int_{-\infty}^{\infty} \int_0^\infty \phi(t) \, G(x,t;y,s)dt \, dW_{ys}$$

$$= -\tilde{V}(x,\phi).$$

Q.E.D.

References

1. R. CAIROLI and J.B. WALSH. Stochastic integrals in the plane. *Acta Math. 134* (1975), 111-183.

2. D.A. DAWSON. Stochastic evolution equations and related measure processes. *J. Multiv. Anal. 3* (1975), 1-52.

3. A. FRIEDMAN, *Partial Differential Equations of Parabolic Type*. Prentice Hall, Englewood Cliffs, 1964.

4. A. GARSIA. Continuity properties of Gaussian processes with multi-dimensional time parameter. *Proc. 6th Berkeley Symp. Math. Stat. Prob.*, Vol. II, 1970-71, 369-374.

5. I.M. GELFAND and N. YA VILENKIN. *Generalized Functions, 4*, Academic

Press, New York, 1964.

6. P. GREINER. An asymptotic expansion for the heat equation. *Arch. Ratl. Mech. Anal.*, *41* (1971), 163-218.

7. G. KALLIANPUR and R. WOLPERT. Infinite dimensional stochastic differential equation models for spatially distributed neurons (preprint).

8. J.B. WALSH. A stochastic model of neural response. *Adv. Appl. Prob.*, *13*, 231-281.

9. J.B. WALSH. Convergence and regularity of multiparameter strong martingales. *Z. Warsch. u. Verw. Gebiete, 29* (1974), 109-122.

10. E. WONG and M. ZAKAI. Martingales and stochastic integrals for processes with a multidimensional parameter. *Z. Warsch. u. Verw. Gebiete, 29* (1974), 109-122.

JOHN B. WALSH
Department of Mathematics
University of British Columbia
Vancouver, B.C. V6T 1Y4
CANADA

PROGRESS IN PROBABILITY
AND STATISTICS

Already published

PPS 1 Seminar on Stochastic Processes, 1981
E. Çınlar, K.L. Chung, R.K. Getoor, editors
ISBN 3-7643-3072-4, 248 pages, hardcover

PPS 2 Percolation Theory for Mathematicians
Harry Kesten
ISBN 3-7643-3107-0, 432 pages, hardcover

PPS 3 Branching Processes
S. Asmussen, H. Hering
ISBN 3-7643-3122-4, 472 pages, hardcover

PPS 4 Introduction to Stochastic Integration
K.L. Chung, R.J. Williams
ISBN 0-8176-3117-8
ISBN 3-7643-3117-8, 204 pages, hardcover

PPS 5 Seminar on Stochastic Processes, 1982
E. Çınlar, K.L. Chung, R.K. Getoor, editors
ISBN 0-8176-3131-3
ISBN 3-7643-3131-3, 302 pages, hardcover

PPS 6 Least Absolute Deviation
Peter Bloomfield, W.L. Steiger
ISBN 0-8176-3157-7
ISBN 3-7643-3157-7, 364 pages, hardcover